CONTENTS

Author's Note ix
Foreword by Charles Garrett........ xix

1. **Introduction**........ 1
   European History Timeline 2
   Why Take up Metal Detecting? 13
2. **Detector and Searchcoil Basics** 21
   Early Metal Detectors 21
   Detection Technologies: Single vs. Multiple Frequency 22
   Searchcoil Essentials 27
   All-Metal (Non-Motion) vs. Discrimination (Motion) 40
   Target Identification with Motion Detectors 43
   Signal Processing........ 49
   Description of Metal Detector Features 50
   Purchasing a Metal Detector........ 54
3. **Hunting Tips and Techniques** 57
   How to Use Your Metal Detector 58
   Tips on Proper Scanning 60
   Target ID Tips for Discrimination Detectors 65
   Target Masking and Discrimination Tips 68
   General Hunting Tips........ 76
   Finding Productive Detecting Areas 82
4. **Recovering Your Treasures** 93
   Treasure Recovery Tools........ 95
   Treasure Recovery Techniques 99
   Treasure Recovery with a Pinpointer 102
5. **European Coin Shooting**........ 107
   The Changing Faces of Currency 109
   European Coin Hunting Tips 117
6. **Historic Hoards**........ 129
   Tips on Cache Hunting........ 137
7. **Brooches, Buckles and Buttons**........ 143
8. **Rings, Jewelry and Figurines** 157
9. **Military Artefacts** 173
10. **Bells, Pins, Odds and Ends**........ 189
11. **Sand and Surf Hunting**........ 195
    Detector Basics at the Beach........ 197

Take Advantage of Tides and Treasure Traps 202
12. Prospecting and Specialty Searches 217
Gold Prospecting in Europe 218
Searching for Meteorites 222
Other Special Search Sites in Europe 224
13. Rallies and Clubs 231
Detecting Clubs, Magazines and European Travel 235
UK Rally Experiences 238
14. Cleaning Your Treasures 249
How Should I Clean My Coins and Artefacts? 250
Metal Identification and Cleaning 252
Common Cleaning Methods Described 255
The Kooistra Silver Cleaning Method 257
Final "Polishing-Up" Methods 261
Photographing and Displaying Your Finds 262
15. European Treasure Hunting Laws 265
16. Metal Detectors and Archaeology 279
How Can You Get Involved With History? 282
The Archaeology Process 284
17. Metal Detectorist's Code of Ethics 289

Selected Bibliography 293

AUTHOR'S NOTE

To proclaim that a metal detectorist can make amazing discoveries in Europe is akin to boasting of "finding" a good restaurant in Paris. The obvious facts of history and culture speak for themselves.

When a searcher in the United States discovers a colonial object that is more than two centuries old—or even a more recent Civil War artefact—the find is considered significant. Meanwhile, in Europe metal detectorists are continually discovering artefacts that date back thousands of years. The same ploughed field can—and very often does—yield coins and artefacts from the Iron Age, the Roman Empire and 20th Century wars. Conquerors, rulers, clergy, nobles and just plain citizens have all left traces of their civilizations across Europe for metal detectors to discover.

This book provides a collection of tips, techniques and treasure images from European metal detectorists who have had success in the field. Some have pursued the hobby since its post-World War II infancy while others are relative newcomers who have quickly become genuine enthusiasts. The passion of finding historic items is hard to ignore. Once you have pulled a nearly-2,000-year-old Roman coin from the earth, it's hard not to be excited.

Garrett metal detectors have been sold in Europe for some 30 years, and their numbers continue to increase as the hobby grows in popularity. Garrett's RAM Books publishing division has released dozens of metal detecting guides during this time, but very few have been focused on the European detecting grounds. Henry Tellez, Garrett's vice president of international sales, brought this shortcoming to RAM's attention.

Henry and I embarked on several journeys to visit with metal detector users, dealers and distributors across Europe to learn as much as we possibly could for this detecting handbook. Our goal was a book that would be educational both to novices and experienced hobbyists on the European continent and to those abroad who might wish to visit the area with their detectors.

My background is in marketing and in journalism, but I have developed a love for finding history with metal detectors. I was understandably excited when asked to write a book for Garrett about the European detecting community. My books during the past 15 years have focused on World War II military, frontier warfare and Texas history. Only recently have my interests in metal detecting and history writing combined. This past year, I led groups of detectorists in a successful effort to pinpoint the key locations of an 1839 Indian battle, the largest ever between Texas Rangers and Native Americans. Our artefact recoveries and the history of this Indian battle are chronicled in the recent *Last Stand of the Texas Cherokees*, released by RAM Books.

A few months later, I organized another search team which included Charles Garrett to work hand-in-hand with archaeologists on the battleground of the key engagement of the 1836 Texas Revolution. In addition to searching historic sites for artefacts, I enjoy coin hunting and joining Garrett detectorists wherever we may travel in our marketing efforts.

In order to truly experience this "Old World" metal detecting community, Henry and I spent a good deal of time on the road this past year to collect stories and photos for this book. The history we study in the United States covers a relatively limited period of time compared with the rich culture of Europe, where a detectorist can dig a target most anywhere that might be thousands of years old. My appreciation of this fact increased immeasurably as we attended rallies in Spain, France and England and also joined

Author Steve Moore holds out an early coin he has just recovered from a Belgian farmer's field during one of his trips to Europe to research this book.

detectorists in the Netherlands, Belgium and Italy in the field.

The photos in this book bear testament to the astonishing age and beauty of the artefacts and coins from the continent. Aside from its fabulous finds, the European community offers a cultural experience that is generally much more relaxed than that of North America. The people also lead somewhat healthier lifestyles. For example, a French proverb related by Gilles Cavaillé could be considered sound advice for any culture: "Eat like a king for breakfast, eat like a prince at lunchtime and eat like a pauper for dinner." This advice is well taken for a detectorist, who should prepare his or her body for outdoor exercise with a good morning meal, followed by a replenishing yet healthy lunch. Most of the calories should be burned while walking and digging in the field. Therefore, a sensible evening meal that is not too heavy will make the day's detecting "workout" good for anyone's health.

We covered thousands of miles on trains, planes, buses and automobiles across numerous countries in search of stories and photos to enrich this book. Each trip was its own adventure in missed flights, lost luggage and other challenges. I can say from experience, however, that Americans should have no problems taking metal detectors overseas.

European Metal Detecting Guide is the result of many people coming together to share their knowledge. Foremost on this list is Charles Garrett—a man who is truly a pioneer of today's advanced metal detectors. He has spent decades in the field hunting and testing instruments and technology that his engineers have implemented into Garrett detectors. I must thank him for the many stories he has shared about his European experiences and for the advice he has provided along the way in compiling this book. We also spent several hours culling through one of Mr. Garrett's vaults, which was a special treat as he related stories of his finds.

Senior Vice President and Director of Engineering Bob Podhrasky and Director of Product Development Brent Weaver, the leading engineers for Garrett Metal Detectors, served as my technical advisors for this endeavour. Brent particularly offered much insight on the science of how and why metal detectors and searchcoils operate as they do. Brian McKenzie spent many hours photographing European treasures and creating superb visuals to illustrate the more technical aspects of this manual. I am also grateful for the editorial skills of Hal Dawson and for artist John Lowe's efforts to improve some of this book's design elements.

Sir Robert Marx, the underseas artefact recovery expert who has been knighted by three European countries, was kind enough to share some of his images and help me identify some items discovered by European detectorists. A RAM Books author for many years, Bob has extensive knowledge of worldly artefacts from all time periods thanks to the decades he has spent working shipwrecks and historic civilizations sites that date back to the earliest uses of metal. His home is something of a museum, decorated with earthenware, figurines, jewelry and countless artefacts Bob has

The author tries his luck searching farmland in France at a rally held near Toulouse.

recovered with metal detectors from Old World shipwrecks.

Beyond these technical supporters, dozens of European metal detectorists provided tips and techniques for searching in their homelands. Some revealed their success secrets and some shared photos of their favorite finds. Although many are named only by their first name (per their request), they all deserve recognition for their contributions. Many others shared stories with me either in person or via email.

In no particular order, the key contributors to this book were: Franco Berlingieri of Muntzoeker Metaaldetectors, Gilles Cavaillé of Loisirs Detection, Nigel Ingram of Regton, Leo Kooistra of Kooistra Metal Detectors, Jesus Condom of Orcrom Metal Detectors, Bruno Lallin of Lutece Detection, Albino Bartolini of EB Elettronica, Dmitry Zatsepin of Rei-Soi, Bernard Sobra, Derek Ingram, Mark Hallai, Omar Aissaoui, Regis Najac, Stefano Morsiani of Electronics Company Italy, Adam Rudnicki, Andrea Scardovi, Mark Ickx, Danny Reijnders, Jason Price, Gary Norman, Sergey Chernokryluk, Andrey Shipko, Viktor Tobolsky, Denis Sokolov,

David Booth, Maxim Burmistrov, Hector de Luna, Ron Herbert and Pavel Popov. Successes of such experienced detectorists from all over Europe offer the key to learning how to become a better treasure hunter.

Coin experts and other specialists may detect errors with how this book reports a particular treasure or tip. Rest assured that efforts have been made to make the text as accurate as possible.

Even in the final hours of preparing this book for press there are great discoveries being made in Europe. One just making news demonstrates how finds occur for both veteran and rookie detectorists: the fall 2009 Iron Age jewelry hoard located by 35-year-old David Booth of Scotland.

Employed as the chief game warden for a wildlife safari park near Stirling, Scotland, David had just taken up the new hobby of metal detecting. He purchased a Garrett *ACE 250* and initially spent a little time practising around his house and garden to begin understanding its capabilities. Five days after purchasing his *ACE*, David drove to a field he had decided would make an interesting spot for his first real treasure hunt.

After parking his car, he turned on his detector in its All-Metal (no discrimination) Mode. "I was still learning how to use the machine and thought it was best to dig everything for the first ten hours or so," he related. He had advanced only seven paces from his vehicle and had spent less than five minutes swinging his searchcoil when he heard a signal that would forever change his life.

Since his *ACE 250*'s Target ID indicated this object to be gold, David began to dig carefully with a garden spade. He made a large circle around the spot he had pinpointed and then switched to a hand trowel as he dug closer to the source of his signal. "It was between six and eight inches under the ground," he related. "As I uncovered the first piece of the hoard, I was stunned. I immediately knew that it looked like something important. However, I did wonder if I could possibly be that lucky on my first trip out."

David had indeed found an incredibly rare hoard of five gold treasure pieces that were grouped tightly together—three intact necklaces and two fragments of another. These four golden neck pieces, referred to as "torcs," were of a Southwest French style annular torc design. They are each of a gold and silver alloy with a touch of copper in their mix. These ribbon torcs, believed to be of Scottish or Irish creation, had been twisted from sheet gold with hinge and catch clasps on their ends.

Excited beyond belief, David took the muddy artefacts home and rinsed them carefully. "It was only when I got home and did some research on the Internet that I was fairly sure they were Iron Age," he noted. David locked the ancient treasures in his gun safe for the night and then took them to his safari park the next morning. He emailed photos of the gold necklaces to the National Museum of Scotland (NMS).

Within hours, NMS representatives had arrived at David's door. The museum's principal Iron Age and Roman curator, Dr. Fraser Hunter, would admit that he nearly fell off his chair when he first saw the photos. Experts have thus far estimated the date of David Booth's hoard to be between 300 and 100 BC. The decorative ends of the looped golden torcs have fascinated Scottish historians. Each terminal piece is made from eight golden wires looped together and decorated with thin threads and chains.

This craftsmanship suggests that the metalsmith had learned his trade in the Mediterranean and had combined his style with that of the local customs. Ian Ralston, a professor of archaeology at the University of Edinburgh, suggests that early Scottish tribes and other Iron Age people in Europe may have been more interlinked than previous research had suggested. Subsequent analysis by archaeologists at the site of David's find produced the remnants of a wooden roundhouse, suggesting that this jewelry hoard was perhaps a votive offering to higher powers or had been cached beneath an old homestead during a time of war.

Scotland's Treasure Trove Unit, based at the National Museum of Scotland, investigates all such metal detector finds. Since ancient times, the common law of Scotland has been that such treasure trove belongs to the Crown. Such rare jewelry will certainly

SAFARI
GARRETT

David posed with his Iron Age gold torcs before they were taken by the National Museum of Scotland in Edinburgh. The four necklaces, three of which are fully intact, are believed to have been crafted by a member of a Celtic-speaking tribe between 300 and 100 BC.

These rare necklaces are considered the most significant find in Scotland since 1857, when two other gold torcs were found on farmland in Moray. Early estimates have circulated that David might eventually be rewarded with £1 million or more for his recovery.

Photos courtesy of David Booth.

become a national treasure for the public museums, and David Booth will be compensated with a reward equivalent to the market value of the find. Early speculation has been that his "Stirlingshire Hoard" could fetch more than £1 million (about $1.6 million U.S.)!

David had long fancied buying his own metal detector to take up the hobby. Within a week of so doing, this young father-to-be made the remarkable recovery that left him believing his *ACE 250* was "a pretty good investment." When I asked David for his treasure-seeking tips for other newcomers to the hobby of metal detecting, he said he would advise:

- Plenty of research on suitable areas to detect;
- Ensure you have the landowners' permission to detect;
- Ensure the site is not listed as a Scheduled Ancient Monument or has any other restrictions;
- Report any important finds to a local museum or treasure trove promptly so that the context of the find is not lost;
- Carry a camera to record exactly how important items were found.

This most recent success by detectorist David Booth is similar to stories that continue to be told by fortunate searchers all over Europe. We hope this European detecting guide will offer some helpful tips on selecting the right metal detector for your needs and provide successful hunting advice you can use in the field. I make no personal claims to be an expert treasure hunter…only an enthusiastic journalist and detectorist who is willing to take on new challenges and learn from the veterans everywhere I go.

My appreciation of European history increases with each visit and I hope to return to experience more of the wonderful metal detecting opportunities. It is truly a hunter's world where centuries of heritage can come to life with any given swing of the searchcoil. History found is history that can be preserved and appreciated.

Foreword

I am often asked, "Is there really much treasure left to be found after all these years of people using metal detectors?" This question always make me chuckle, and I quickly reply, "Absolutely!"

Metal detector technology continues to improve. Today's high-end metal detectors have better discrimination ability, faster digital signal processors and improved searchcoils that allow them to search deeper and more selectively in soil conditions of all types—mineralized ground, areas littered with both old and modern iron "junk metal," and in the trying wetted beach salt sand. The farms, fields, cities, parks, battlegrounds, beaches and ocean floors will continue to yield fabulous coins, relics, jewelry, gold nuggets and historical artifacts of all types for centuries to come. In my estimation, we have unearthed no more than 1% of the valuable items that lie hidden below.

Treasure, in fact, is a commodity that continually replaces itself. Every day throughout the world people lose coins, medallions, tokens, jewelry, buttons and personal items of all description that will become someone else's great discovery. In that regard, Europe is a hunting ground like no other. This area is rich with the history of cultures that have clashed for literally thousands of years as rulers, dictators, conquerors and settlers roamed this vast continent.

For decades, I have field tested my metal detectors around the world, including many ventures into the European fields. I made many good friends during these outings and they have shared their tips and techniques for productive hunting. I have found ancient Roman and Greek coins and have recovered military ar-

(Left) Charles Garrett searches around Glamis Castle in Angus, Scotland in 1986. He was asked by the Countess of Strathmore to search for several items lost on the grounds.

(Right) Garrett continues to field test his metal detectors. He participated in searching a Texas Revolution battlefield in 2009 while working with state archaeologists.

tifacts from French, German, Scottish, English, Italian and many other cultures. My son Vaughan once accompanied me on a two-week excursion through a half dozen European countries. I still recall his thrill of finding his first two Roman coins, which were nearly one thousand years old.

European Metal Detecting Guide is filled with photos of such great treasures, many of which predate the time of Jesus Christ on this earth. The early chapters provide an excellent perspective on how metal detectors and searchcoils work, as well as useful information to help you decide which metal detector is right for your needs. Throughout the text, European detectorists share some of their favorite finds and offer advice on the techniques that have helped them become more successful in their searches.

I would strongly advise those new to the sport to join a treasure club in their area to fully benefit from this hobby. You will often establish lifelong friendships with people who share common interests. European metal detector rallies are held each year in places that might otherwise not be available for you to search. As you recover each metallic object of the Old World, take a moment to consider how it was lost and the history that object portrays. Modern metal detectors are finding relics that archaeologists have only dreamed of, for these computerized marvels have "eyes" that can see far beneath the soil.

European detectorists must adapt to the ever-changing antiquities laws that restrict hunting freely in certain areas of great historic importance. I urge you to study the antiquity laws in your country and "know before you go" into the field. The United Kingdom has enacted legislation in recent years that actually encourages detectorists to report their important discoveries. The British Museum often purchases significant finds at fair market value, a process that has greatly increased the percentage of recoveries that are reported. Because other European countries are far more strict on allowing the use of metal detectors, be mindful of the restrictions in each country and province as you travel with your detector. I urge you to read and follow the European metal detecting code of ethics presented in this book. Responsible recovery of ancient items opens the door for continued harmony between detectorists, historians and archaeologists.

Through the years, I have talked with countless European detectorists who have shared their secrets of combining good research and perseverance. Jean Ward of Mansfield, England, for example, studied aerial photographs before going to work in the Nottingham potato fields. Her early finds were good—Victorian coins, George II, George III, a hammered Henry III silver coin, musket and pistol balls—but she was on a quest to find Roman coins in these fields. Jean continued to study one particular field more carefully and she finally selected a particular location. Her hard work was rewarded with a treasure trove of Roman coins—29 denarii, mostly from the 1st and 2nd centuries AD—and her prize was purchased for display in a local museum. During the course of her searches, she was also able to assist the local farmer by recovering a part he had lost from his potato machine while plowing.

Arnold Anderson took his Garrett detector along on a visit to his daughter's home in Europe, where they visited a 29-room chateau in the French Alps whose construction began in the 15th century. On the chateau's grounds, Arnold found thimbles, World War

II projectiles, and more than 100 coins from nine different countries (including France, Switzerland and Italy) that dated back as early as 1610.

My friend Franco Berlingieri of Belgium has used Garrett detectors to discover tens of thousands of European coins, including rare Greek, Roman and Celtic coinage. He willingly offered his time, techniques and treasure images to this book.

I would like to thank all of the other detectorists who chose to share photos and stories of their discoveries with Steve Moore and Henry Tellez for this book. The information presented here is informative for both the seasoned detectorists and the new-comers to our hobby. I am pleased every time I read testimonies from happy Europeans who have dug ancient items with their Garrett metal detectors. I advise you to follow the law, conduct proper research to find those special "hot spots," and above all, have fun as your quest continues.

I hope to see you in the field,

Charles Garrett

Garland, Texas

CHAPTER 1

INTRODUCTION

The hobby of metal detecting has been growing in popularity at a rapid pace during the past several decades. More and more people of all ages, both men and women, have learned how much fun it is to recover treasure with modern electronic machines. This sport offers avid hobbyists the chance to enjoy the fresh air of the outdoors while also providing exercise, relaxation and a mental escape from the grinds of daily routines. With a little luck, patience and research, dedicated detectorists also afford themselves the opportunity to make a truly remarkable or historic discovery.

In the United States, hobby detectorists are thrilled to find an "old" Civil War, Revolutionary War or colonial artefact dating 150 to 300 years in age. European detectorists, on the other hand, can unearth metallic items and coins that date back to the Bronze Age, more than 4,000 years ago. Any worries that treasure fields would be picked clean of riches have been laid aside during the past 20 years as metal detectors have become more and more technologically advanced. Charles Garrett, a pioneering engineer of today's modern metal detector and an avid treasure hunter himself, estimates that more than 95 percent of the world's treasures remain to be discovered in the fields, streams, oceans, mountains, countrysides and settlements of Mother Earth. Add to this estimate the undeniable fact that millions of new coins and other metallic items are lost or discarded across the world each year, providing a vast treasure trove that continues to replenish itself daily.

Today, the 27 countries of the European Union (EU) represent one of the largest single developed markets in the world with more than 450 million consumers. The countries, or member states, currently comprising the EU are Austria, Belgium, Bulgaria, Cyprus, Czech Republic, Denmark, Estonia, Finland, France, Germany, Greece, Hungary, Ireland, Italy, Latvia, Lithuania, Luxembourg, Malta, Netherlands, Poland, Portugal, Romania, Slovakia, Slovenia, Spain, Sweden and the United Kingdom.

The metal detecting hobby continues to grow in popularity within Europe and U.S. hobbyists travel overseas for the chance to recover ancient coins. Many detectorists have formed societies and clubs dedicated to enhancing their treasure-seeking hobby. Some of these groups unite with professional archaeological projects to lend their electronic expertise in unearthing valuable historic artefacts that contribute to our understanding of cultures and societies of past wars, settlements and ancient civilizations.

European History Timeline

History lovers know that Europe is perhaps the world's richest metal detecting environment for ancient artefacts and coins. It is important to understand the basics of the historical timeline of Europe in order to begin to estimate how old an item is that you recover with your detector.

The first evidence of man's existence in Europe dates back to around 35,000 BC, during the European Paleolithic period. The Neolithic period, around 7000 BC, marked the appearance of settlements, agriculture and domesticated livestock. The preferred metal for tools and weapons of this more civilized period was bronze, and historians have labeled this the Bronze Age.

The Roman Republic, a phase of the ancient Roman civilization characterized by a republican form of government, was established around 510 BC with the overthrow of the previous monarchy. During their Republic era, Romans expanded territo-

EUROPEAN HISTORY TIMELINE FOR DATING ARTEFACTS

The following chronology is a simplified list of some commonly regarded European historical eras. Note that some time periods have overlapping dates because of the vast area of human population in what is now Europe.

Time	Period	Brief Description
7000—3000 BC	Neolithic Period	Settlements, agriculture and domesticated lifestock appear.
3300—1200 BC	Bronze Age	Tools, weapons and early metalworkers utilize bronze, which is smelted of copper and tin from ores.
1200 BC—400 AD	Iron Age	Cutting tools and weapons were largely made of iron or steel during this long period of European history. Lydia (present Turkey) produces first coins, made of electrum (an alloy of gold and silver).
1100—146 BC	Ancient Greece	This period includes the *Greek Dark Ages* (1100–750 BC), the *Archaic Period* (750–480 BC), the *Classical Period* (500–323 BC) and the *Hellenistic Period* (323–146 BC). The Hellenistic Period began with the death of Alexander the Great and ended with the Roman conquest of Greece in 146 BC. By this time, Greek culture and power expanded into the near and middle east.
800—500 BC	Classic Era/Classical Antiquity Period	Period of cultural history during which the Greek and Roman arts flourished.
509—27 BC	Roman Republic	This period began with the overthrow of the Roman monarchy in 509 BC. The new republican form of government lasted more than 450 years until it collapsed due to a series of civil wars.
27 BC—476 AD	Roman Empire	This period of autocratic rule over much of Europe began with the first emperor, Augustus, and continued for more than four centuries until, weakened by invasions, the last Roman emperor was removed from power.
448—751 AD	Merovingian Period (France)	Transition period in France after the collapse of the Roman Empire during which a dynasty of Merovingian kings ruled the Frankish

EUROPEAN HISTORY TIMELINE FOR DATING ARTEFACTS (cont.)

		kingdom before the time of Charlamagne and the Holy Roman Empire.
476—1350 AD	Middle Ages	A period of European history called the Dark Ages (or Medieval Period) considered to have lasted from the end of the Roman Empire until the rise of national monarchies.
711—1100 AD	Iberian Peninsula	Moors dominated this area of modern-day Spain and Portugal.
793—1171 AD	Viking Age	During this Scandinavian (Norway, Sweden and Denmark) and Northern European history period, Vikings (or Norsemen) explored Europe by oceans and rivers through trade and warfare.
1066—1200 AD	Norman Period	The Norman conquest of England by William the Conqueror linked the island more closely with continental Europe.
1350—1600 AD	Renaissance	This French term for "rebirth" describes the artistic and economic changes that occurred in Europe from the 14th through the 16th centuries.
1517—1800 AD	Enlightenment	This period marked an intellectual "awakening" movement beginning in 16th-century Europe which led the world toward progress. This movement provided a framework for the French Revolution (1789-1794) and the coming of Napoleon.
1750—1900 AD	Industrial Revolution	A period of social and technological changes marked primarily by significant manufacturing advances.

rial control from Italy over the Mediterranean area and into western Europe. Rome grew to dominate North Africa, Iberia, Greece, the British Isles and what is now France.

The end of this 450-year Roman Republic period was followed in 27 BC by the period known as the Roman Empire. This term is used to describe the Roman state during and after the time of the first emperor, Augustus. The greatest extent of Roman control

Ancient Greek ruins which still stand in Paestum (south of Naples) precede even the Roman Empire that once controlled much of Europe for about 450 years. Metal detectorists find Roman coins and other artefacts in regions far removed from Italy. *Courtesy of Ted Johnson.*

of European territories was reached around 150 AD. The Romans spread literature, engineering and architecture throughout Europe and in the process of their expansion they also spread countless coins and artefacts for today's metal detectorists to find.

Christianity later spread over the area after emperor Constantine legalized the Christian religion in the fourth century AD. By 476 AD, the last Roman emperor in the west (Romulus Augustus) had been removed from power, although southeastern Europe and parts of the Mediterranean would remain under the rule of the Roman Empire from Constantinople.

Emperor Justinian then sent Roman armies to restore imperial rule to many parts of the Mediterranean, but this expansion began to erode late in the sixth century. As Emperor Constantinople's hold on western territories faltered, wars were fought as more Germanic peoples invaded and established their own kingdoms within Europe. The eastern Mediterranean territories that remained largely in the hands of the Christian emperor in Constantinople through the sixth century were later labeled as the Byzantine Empire.

Celts (pronounced "Kelts") is a term used to describe any of the European peoples who spoke a Celtic language. The earliest Celts were a non-literate culture whose history was preserved through oral tradition. The Greeks and Romans were impressed by the powerful appearance and bold dress of the Celtic people, who were technically and artistically advanced metal craftsmen. In the earliest days of European history, the Celts then living in present France were known as Gauls to the Romans. Celtic people occupied land in modern day Ireland, Scotland, Wales, England, Western Europe, Eastern Europe, Northern Italy, Greece and Spain.

Because Celts were so widely spread throughout Europe, coins and metallic artefacts of their culture are found by metal detectorists today in many countries. Celtic coins were banned from circulation in 61 AD as an act of retribution by the Romans after the Celts had staged a rebellion. Many Celts were enlisted in the Roman Army in the late days of the Roman Republic. Detectorists often find Roman and Celtic artefacts in the same field.

In England, Anglo-Saxon immigration increased significantly as the Roman Empire collapsed. Saxon kingdoms were established in the fifth century until the Norman Conquest of England in 1066.

The vast history of Europe simply cannot be presented in a text such as this, but the point is that the Roman Empire and the Celtic culture once sprawled over a vast area of present Europe. What does this mean? Simply put, Roman and Celtic treasures can be found from Italy all the way north and west to the Netherlands and even on the British Isles.

In Europe, you can literally find coins and artefacts spanning thousands of years and from various empires in the same field. There are countless settlements in France, Spain, Holland, and countries all across Europe where Celtic, Greek and Roman people once lived thousands of years ago. Detectorists scanning through a newly ploughed field are just as likely to turn up a fine Roman stater as they are to find a modern euro.

This collection of recovered treasures illustrates the range of history that can be found in a single field in Europe. More modern items from the past hundred years are often mixed in close proximity to items from the earliest Roman and Celtic history. The small bronze coin at the very top of this iron pot is a Celtic coin more than 2,000 years old. Directly below it on the second through fifth rows of items are copper and bronze coins and artefacts from the Roman empire that range in dates from about 2,000 years ago to the 350 AD period. The bottom two coins on this pot are an 1887 large Belgian coin and a 1938 German coin from the Nazi era. Among the other early pieces of history in this photo are fibulas, a spoon, a ring from a pot holder, and a large bronze button. All of these variously dated items were found over an eight-year period with Garrett detectors in Belgium **at the same ancient Groot Loon town site.** Every several centimeters of depth can produce a new century of artefacts in some cases.

(Right) Fibula, ring and other artefacts from above shown in detail.

European farmland is an excellent site for hunting coins and artefacts. Many detectorists seek permission from their farming friends to search fields after a fresh ploughing (and before crops are planted) to discover what the farm equipment has stirred up from thousands of years ago.

Why were so many coins lost in the fields in Europe? There is no single answer to this question but here are several offered by various Europeans:

• *People spent time in the fields as a means of survival.* Hunting, fishing and farming were three of the key means of subsistence in the early days. Many people had to work in the fields to maintain a village. If an early village contained a thousand or more people, hundreds of them might be engaged during the year to help plant, till and harvest crops.

Long ago, Romans and Saxons buried their dead in the field in the vicinity of their village. Some of the wealthier Saxons have been found by detectorists. Ancient Egyptians buried pharaohs with elaborate treasures. Early Saxon nobles were also buried with elaborate brooches or other treasures. In other early Roman burials, a coin was placed into the deceased person's mouth to pay for passage to the next world. A rich man might have had a gold coin placed in his mouth.

Such burials are mentioned because European detectorists do occasionally stumble upon an ancient gravesite while digging a deep target reported by their detector. In the event you should find one, report it to your local coroner or police station immediately. Such grave sites are treated like crime scenes because human remains are involved.

- *Hoards or caches were buried by people.* In the Old World, people were paid in coins. There were no paper bills. People often buried some of their coins in hoards to protect them from robbery before there were banks. Perhaps the house was destroyed by a fire or natural disaster or maybe the person died without telling anyone where it was buried.

Some people may have been quickly forced from their homes by another conquering force. That person might hope to one day return for their money but often times this was never possible. There were no banks in Roman or Medieval times, so coins and other valuables were simply buried away from the home near an identifiable natural marker. In some cases, people might have simply forgotten where they had cached their hoards or they became sick or were killed before telling someone else.

- *Old coins and artefacts are continually moved about by agricultural efforts.* Ploughs have tilled up the land in Europe for thousands of years and these same ploughs—whether pulled by oxen or churning behind modern tractors—move coins and artefacts from their original point of rest. Tractors can go as deep as a meter or more, depending upon the type of plough being used and the condition of the local soil. Some of the soil is so rich and so soft because it has been farmed for countless generations.

Experienced treasure hunters know that the same well-hunted field can still continue to produce finds year after year. Some such hunters simply return to their favorite hot spots after a farmer has ploughed his land in preparation for the planting season. A fresh ploughing produces a fresh hunting field.

Today's metal detectors continue to become more sophisticated with electronics that allow them to search deeper and to better discriminate out unwanted "junk metal" targets. Newer detectors have microprocessors or even digital signal processors that can be adapted to the ground mineralization and to the swing speed of the hunter. In many cases, the hunter once had to swing his or her older machine at a controlled, slower speed to make certain the equipment properly processed treasure signals.

Today's higher-end machines mean that detectorists can swing the searchcoil at their preferred speed and still get accurate target reporting. The experienced hunter, therefore, can cover more ground in less time, hunt deeper than before, ignore more trash items, and recover targets much quicker than in past years. In short, an old field that has been newly ploughed is often worth the effort of another search—particularly if a hobbyist has "stepped up" to a better metal detector.

• *People threw coins into the field for good luck.* Farmers in centuries past often threw coins into their field if they had a particularly good crop season. People have long made

The Trevi Fountain in Rome is a popular tourist draw where visitors toss in their coins for good luck. The largest Baroque fountain in the city, the current structure was completed in 1762 to mark the terminal point of one of the city's early aqueducts.

wishes by throwing coins into a fountain or lake. In the early days before fountains, good luck pieces were tossed into streams, creeks or bodies of water that may have long since dried up or changed course. The tradition of tossing coins into fountains or bodies of water is an age-old custom. Even today, the Trevi Fountain in Rome is a popular destination for coin tossers who believe they will find happiness in their love life by flinging pocket change into its water. Approximately 3,000 Euros are tossed into Trevi Fountain daily and the money has been used to subsidize a supermarket for Rome's needy.

- *The coins were so heavy that they would break through the bags the workers carried.* People working in the field to harvest crops walked all over the fields, picking the crop. Certainly, they lost some coins in the process. Older coins, due to their heavy weight, could rip a bag. People might have been travelling through this area and simply lost the coins.
- *The field might have been a farmer's market at one time.* Sometimes in an area where trails or roads converged, local farmers would have merchant marts or vegetable and fruit stands to sell their goods. In such areas of commerce many coins would be lost. These rural farmers markets have often been replaced over the past centuries as the world has become more "civilized" and more

Joseph D. of Russia made these recoveries, ranging in age from World War II era to the Middle Ages, with his Garrett *ACE 250* metal detector from the same field. *Image courtesy of Dmitry Zatsepin of Rei-Soi.*

Europe is filled with old ruins and sites of ancient villages and marketplaces—any of which can yield exciting recoveries with metal detectors.

people buy their goods in the city. What may once have thrived as a local shopping mart in the country might simply be another few acres of fruits and veggies in the countryside today.

• During the different empires that ruled the European world, coins changed their value. Sometimes a new emperor came into power and the old coins of another emperor weren't worth much. Some *people simply discarded their old worthless small coins* in their rubbish dump. The good, organic trash (food scraps, etc.) were collected in compost piles to be redistributed as fertilizer on the fields. Such decomposed materials were spread out over the fields.

Some coins were "double-hit" or restamped. When the government or emperor changed hands, the new regime could recall coins to be restamped with the new emperor's face. In modern times, countries such as Mexico and the U.S. have simply changed the look of their older currency. In the 16th and 17th centuries in

European detectorists have the opportunity to find historic coins that are works of art.

Europe, citizens were asked to bring in their coins to be restamped when there was a change in power. Some were restamped with a higher face value during a financial crisis of the 1600s.

Why Take Up Metal Detecting?

• *You can enjoy it wherever you go in Europe.* Great recoveries can be made in almost every corner of the continent. Take your metal detector when you go camping and search around the area. Be sure to study the local laws concerning metal detector use before you travel. The laws vary widely across Europe and are in a state of change due to pressures to protect antiquities. (Chapter 15 offers basic information on the laws within each European nation.)

• *It's something you can do with your children that you can all enjoy.* By the age of five, kids can operate basic metal detectors such as

Metal detectorists in Europe enjoy the sport of finding exciting old coins and artefacts. The hobby offers the benefit of good exercise in the outdoors and a better appreciation of history.

Garrett's *ACE 150*. It is generally better to start them with such a machine that has limited modes and is light in weight. Your children will be entertained even with finding modern Euros. Metal detecting thus becomes a hobby where you can take the kids along and spend time with them, challenging them to games of who can find the best treasure.

- *It is a excellent way to relieve stress.* You search for the good coin just like a fisherman goes out always hoping to land "the big one." More often than not, the fisherman returns with only small catches and sometimes with nothing at all. The treasure hunter is the same: some days he or she might net some great "catches," while other days he or she returns home with nothing at all. Regardless of the success or lack of success, metal detecting enables a person to unwind from the real world's troubles. It's the thrill of the chase that excites as much as the success of the catch.

These assorted items were found in a European field in one afternoon's search made by the author and several other detectorist in April 2009. Their finds include uncleaned coins of various ages, 1700s-era French musket balls, weights and an early square nail.

One detectorist encountered at a European rally admitted that he was a recovering alcoholic. He said that metal detecting gave him an escape from wanting to drink because of how much fun he has in the field. So, in addition to the exercise, there is also the occupation of time for those who might otherwise look for trouble.

• *Make new friends with common interests.* The best way to truly enjoy metal detecting is to join a metal detector club in your area. Your individual hobby of treasure hunting can became a group hobby thanks to your new social organization. Many detectorists find that their new best friends are the people with similar hunting interests.

Club outings and regional rallies add an extra level of social interaction. European rallies often provide the chance to hunt a special historic area that would not otherwise be open to the average person. Such rallies sometimes offer nice prizes for the best finds, most finds, most metallic trash removed from the field and various other categories.

EUROPEAN HISTORY: Old and Older

(Above) Mark Hallai of Belgium displays his first two significant finds made with his Garrett *ACE 250* in the same area.

(Right) Mark holds out his 1792 Louis XVI silver French coin (worth about $1,000 euros) and a gold Roman ring, which has an estimated age of about 1,000 AD.

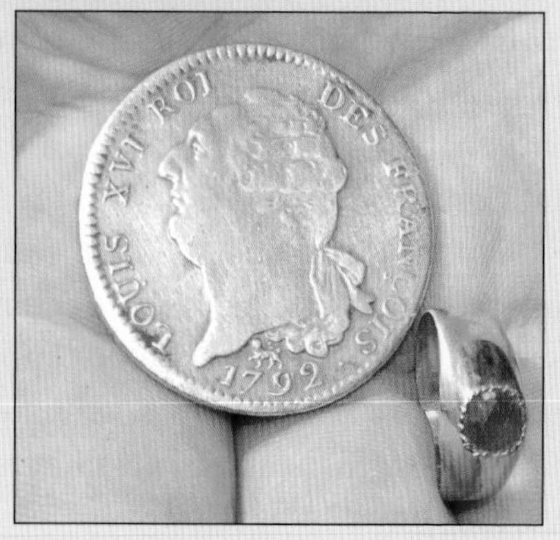

Jan S. of the Czech Republic hunts for historic artefacts in his country using his Garrett *ACE* metal detectors *(seen above)* and the *GTAx 550* he has since acquired.

• *Take up a hobby that can become a profession.* Some children who once enjoyed digging contemporary coins in the park with a metal detector graduate as archaeologists with a special understanding of the power of metal detectors. The term "treasure hunter" is despised by some old-school historians who criticize the recreational detectorists for disturbing potential sites of national historic interest. The more generic term of "metal detectorist" is thus favored in some circles.

• *Detecting is a hobby that can improve your health.* It might sound odd at first, but think about it. This sport requires you (in most cases) to be outdoors where you can enjoy the fresh air and sunshine during good weather. You are constantly walking, which builds leg strength, stamina and is good for your heart. Although your metal detector might be fairly light you will strength your arm or arms by swinging the machine all day.

Finally, if you are in a productive area you will do a lot of stooping and digging. The soil in many areas can make the process of digging to the proper depth quite labor intensive.

Although all of this sun, fresh air and exercise is good, you should prepare yourself for the elements you may face in order to enjoy your day's hunting more. Bring extra clothing for warmth or a rain slicker for inclement weather to keep your body warm. In sunny weather, use plenty of sunscreen or long sleeved garments with ventilation to allow the heat to escape. Always bring plenty of water to keep your body hydrated and bring along a lunch and energy snacks to refresh yourself during long days in the field.

Today, many amateur (advocational) archaeologists take to the field every day to help uncover significant pieces of man's past. Regulations are increasingly adopted to draw the line between what is acceptable conduct for "treasure hunters" and what is not. Some countries have exceedingly stringent laws while others have literally forbidden the use of metal detectors by anyone not carrying credentials.

The challenge that European detectorists now face is how to comply with the local regulations while appealing for the rights to hunt fairly within less constrictive boundaries. Those who follow the general metal detecting code of ethics will help demonstrate to the historians that the worldwide pastime of metal detecting can be enjoyed without disrupting the quest to preserve history.

Metal Detectors Open Doors to History in Russia

Pavel Popov is pleased that the sport of metal detecting has become popular in his Russian homeland. "This was facilitated by a loosening of the laws in relation to the found objects," he explains. "In even the very remote areas of our country there are now people with metal detectors."

Pavel purchased a Garrett *ACE 250* with an additional larger diameter searchcoil to hunt for deep objects. He eagerly set out for the fields this past spring "with a dose of adrenaline like a hunting dog" ready to pursue his targets. "The hot sun of spring was creating a rising steam as it melted the winter snow into a babbling brook that was gaining force.

"Even the smell of the warmed pine needles brings to mind emotions that are hard to forget. I think in every adult male there is some part of the boy who, a few decades ago, dreamed of finding treasure on a deserted island or at least finding a treasure map.

"Metal detecting is a way to find real treasure today. Every subsequent discovery ignites your enthusiasm and anticipation for your next dig. A metal detectorist is not slowed down by such interferences as bad weather, heat or high grass."

Pavel has had good fortune with his finds during 2009. In addition to ancient coins, he has had the good fortune of detecting a Scythian bronze spear point that dates back to the third or fourth century BC. Scythians were pre-Common Era nomadic tribes who traveled extensively throughout Russia and Europe on horseback.

"When I found my first spear tip, I was terribly pleased with the finding," Pavel admits. His ability to find such ancient items is not all coincidental. "Determining such places of settlement takes much preparation to study old maps and learn from the descriptions given by old-timers.

"Struck by the technology with which this javelin point was made, I talked with the director of the local history museum. I learned that the Scythians in our land lived just in the

Pavel Popov uses his metal detector to help preserve ancient Russian history. Some of his earliest finds, such as this Scythian spear point *(left)* are on display in his local museum.

south and the west. I passed along my spear tip and several other finds to our museum since they were of great interest to archaeology. That's how I started as a local historian and how I contribute in my own small way to the study of the history of our country.

"Metal detecting in Russia has been exposed to not only young people but even to those of more venerable ages. We draw up teams to search for early artifacts to preserve our history. Even some of the wives of our esteemed colleagues walk the fields with us and often find more than their husbands."

Pavel is less than pleased when he and his companions come across fields where careless detectorists have left their excavation holes without bothering to fill them in. "The next time, people will not be allowed to go to this place to search," he says. "My hope is that more dedicated people will join our metal detecting hobby.

"We should not do things that will cause detectorists to be viewed in a negative attitude by archaeologists. We should show them that we are not bad people. We may not have the same education that the archaeologists have but we have a common goal: to help preserve our country's history."

CHAPTER 2

DETECTOR AND SEARCHCOIL BASICS

Early Metal Detectors

Scientists and engineers began experimenting with electrical theory in the late 1800s to create a machine that could pinpoint metal in order to locate various ore-bearing rocks. Among the early scientists working on what were the forerunners of modern metal detectors was Scottish physicist Alexander Graham Bell, better known as the inventor of the telephone. Bell was working on an electrical induction balance device for locating metals in 1881, when U.S. President James Garfield was severely wounded in an assassination attempt. He was called upon to use his new device to locate a bullet which had lodged in the President's chest. Dr. Bell's attempts to locate the bullet, however, were unsuccessful before President Garfield died.

The popularity of hobby metal detecting surged following World War II when thousands of surplus military metal detectors were put on the market in America. Prior to 1945, the military had used detectors in Europe for minesweeping purposes. Lieutenant Josef Stanislaw Kosacki, a Polish officer attached to a unit stationed in Scotland, had refined earlier metal detector designs into a more practical unit. Although his creation was heavy, ran on vacuum tubes and needed separate battery packs, this mine detector was used to clear the minefields of retreating Germans during the Second Battle of El Alamein. The Allies also used military mine detectors to clear the beaches during the invasions of Normandy,

Sicily and Italy. Following the war, former American soldiers who had been trained with these detectors purchased surplus units to hunt for coins as well as American Revolutionary War and Civil War artefacts. By the 1950s several companies were manufacturing somewhat more modern metal detectors for hobby enthusiasts.

The number of different types of metal detectors offered on the market today and the claims of their manufacturers can make the purchase decision a daunting task. It is good to understand the essential components of a metal detector and the advantages and disadvantages of the different technologies involved. Key among the components worthy of discussion here are: **detection technologies, searchcoils, target identification** and **signal processing**.

Detection Technologies: Single versus Multiple Frequency

A metal detector transmits magnetic energy into the ground and senses distortion in the magnetic field caused by the presence of a metal object. The frequency content, temporal form and amplitude of this magnetic energy can affect detection capabilities and overall performance characteristics.

The two primary metal detection technologies used in today's detectors are **Single-Frequency** (also known as Continuous Wave or VLF) and **Multiple-Frequency** (examples include Pulse Induction and Dual Frequency). Since each technology has its own detection characteristics, understanding these will enable the purchaser to choose the right detector for his or her treasure hunting needs.

Single Frequency Detection: Most commonly-used modern metal detectors are **VLF** (Very Low Frequency) models whose single frequency generally ranges between 3 kHz and 30 kHz. Such single-frequency detectors are the most sensitive in lightly mineralised soils where they are able to offer the most accurate and reliable target ID and discrimination.

Continuous Wave may be a more descriptive term for Single Frequency technology because it works as the name suggests. The magnetic field generated via Single Frequency technology resembles the continuous flow of waves onto a beach, which results from the continual transmission of energy from the coil. As a result, all of the detector's magnetic energy is focused at a single, powerful frequency *(see illustration on page 25)*. For the vast majority of hobbyists, the best detector will use Single Frequency (VLF) technology because it offers greater depth capabilities, better discrimination and enhanced target ID under those soil conditions where most treasure hunting occurs.

Single Frequency detectors also offer the greatest number of features—such as Target ID, Tone ID, Notch Discrimination and Imaging (a Garrett exclusive). Many of today's Single Frequency detectors offer both All-Metal and Discrimination modes. Such machines can be used for coin, cache and artefact hunting as well as prospecting on land and in streams. These detectors are more

Examples of multiple frequency/PI (Pulse Induction) and single frequency, VLF (Very Low Frequency), metal detectors manufactured by Garrett. Moving from left to right are two PI models and three VLF models: the *Infinium LS*, the *Sea Hunter Mark II*, the *GTI 2500*, the *GTP 1350* and the *ACE 250*.

stable and less susceptible to electrical interference because of their focused bandwidth. Single Frequency machines are also very energy efficient, thus providing longer battery life. They are the most commonly purchased detectors because they are the most appropriate for detecting the broadest range of targets in most soils.

The Single Frequency machine, however, can suffer performance loss in saltwater and over heavily mineralised soils. Large mineral deposits generate strong signals that can distort a target's true signal, resulting in loss of depth and target ID accuracy. For example, a detector using Single Frequency technology may detect a target at 25 centimeters in lightly mineralised soil but may detect the same target at only 10 to 15 centimeters in heavily mineralised soil. Though the detector's magnetic field is still penetrating at least 25 centimeters into the mineralised soil, its ability to recognize the presence of a target at 25 centimeters is limited by the vast amounts of minerals present in the soil.

Single Frequency models offer better depth penetration and ground elimination than their **TR** (Transmitter/Receiver) or **BFO** (Beat Frequency Oscillator) predecessors. Some detectorists still employ detectors with TR technology for coin hunting but their general lack of Discrimination have made them less desirable for most searchers.

Multiple Frequency (PI) Detection: Where Single Frequency technology suffers (i.e. in saltwater environments and highly mineralised soils) Multiple Frequency technology prevails. The best Multiple Frequency technology is found in many of today's specialty metal detectors.

Though technically interchangeable with Multiple Frequency, the term Pulse Induction (PI) may offer a clearer idea of how the spectrum of multiple frequencies is typically generated and utilized. To illustrate, the repetitive transmission of magnetic pulses is similar to the recurring transmission of sonar "pings" (each ping being analogous to an individual magnetic pulse). The spectrum, or frequency content, of the sonar ping contains numerous frequencies called harmonics. This describes how the brief ping can

SINGLE VERSUS MULTIPLE FREQUENCY DETECTORS

ADVANTAGES OF SINGLE FREQUENCY (VLF) DETECTORS

- More sensitive in the most common environments
- Most accurate and reliable target ID and discrimination
- Very energy efficient (long battery life)
- Less susceptible to external noise and interference
- Most appropriate technology for detecting the broadest range of targets over the most common ground conditions
- Majority of hobby detectors use single frequency (VLF) technology

DISADVANTAGES OF SINGLE FREQUENCY (VLF) DETECTORS

- Loss of detection depth and target ID accuracy in saltwater environments and over heavily mineralised soils

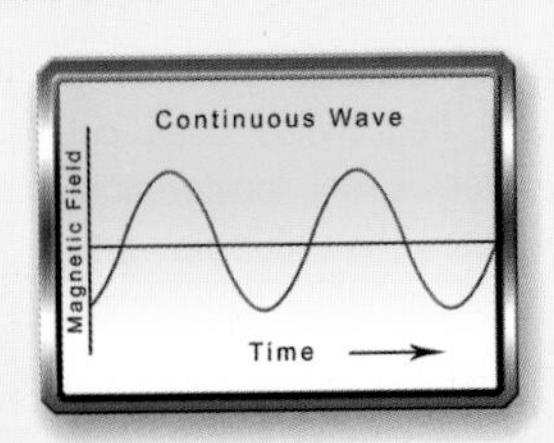

(Above) With a Continuous Wave detector, the magnetic field is continually alternating from positive to negative, thousands of times per second.

(Below) All of the magnetic energy from a Continuous Wave (VLF) detector is focused at one frequency, hence the term "Single Frequency."

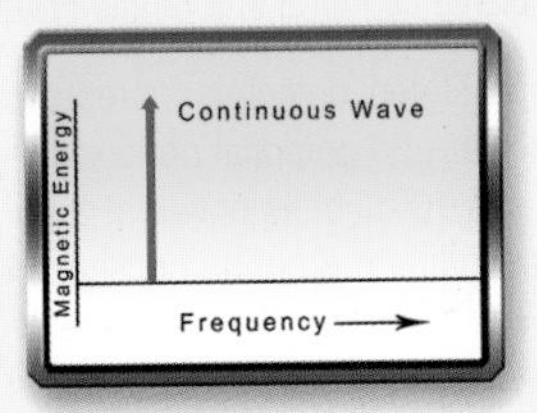

ADVANTAGES OF MULTIPLE FREQUENCY (PI) DETECTORS

- Provide the best results over mineralised ground and in saltwater environments due to its ability to ignore minerals by virtue of their pulse characteristics

DISADVANTAGES OF MULTIPLE FREQUENCY (PI) DETECTORS

- Target ID accuracy suffers, primarily in the region of iron identification
- Size and depth measurement and advanced discrimination modes suffer or are not available
- Inherently less sensitive than Single Frequency
- Typically less energy efficient, requiring many models to use larger batteries
- More susceptible to external noise and interference

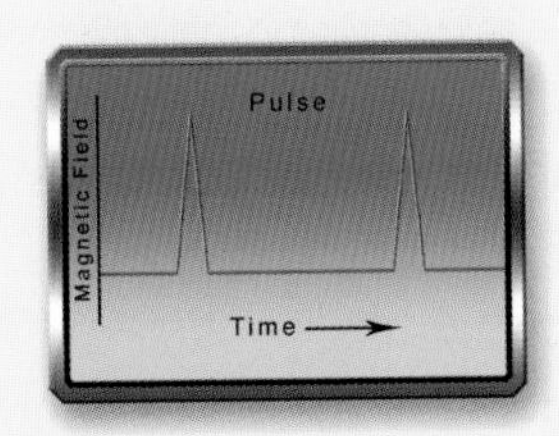

(Above) With a Pulse Induction (PI) detector, the magnetic field is produced by very brief, repetitive spikes or pulses.

(Below) The magnetic energy of a Pulse Induction detector is spread out among many different frequency harmonies, hence the term "Multiple Frequency."

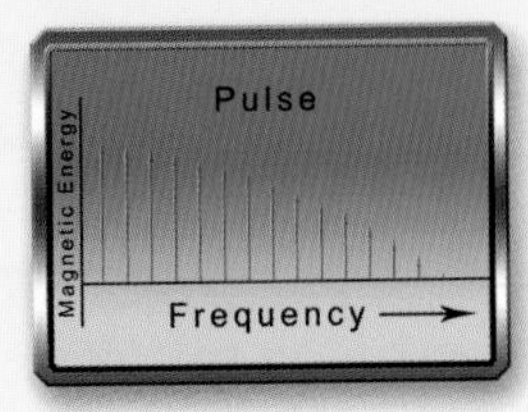

FREQUENCY VERSUS POWER AND PERFORMANCE

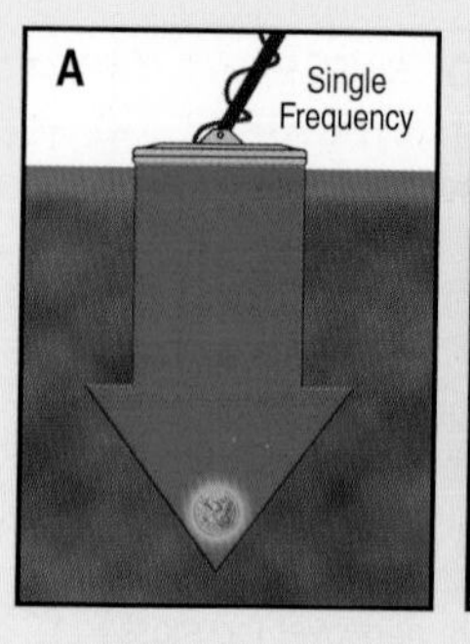

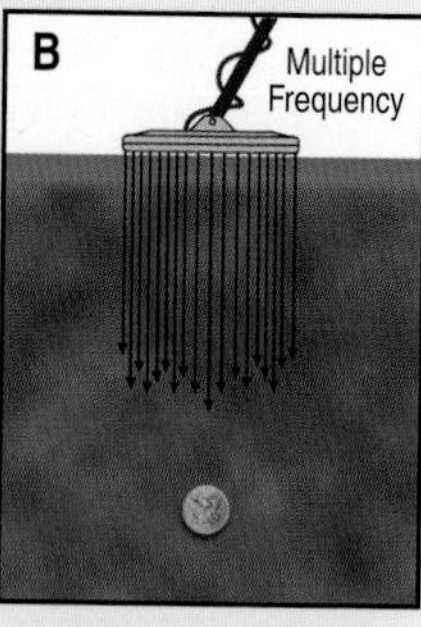

A metal detector transmitting multiple frequencies does not guarantee it will locate more treasures. The Multiple Frequency illustration (B) depicts 18 frequencies (indicated by 18 arrows), which might make you believe you have 18 times the power and performance than that of a Single Frequency detector. In reality, what each arrow in illustration (B) represents is a frequency component with only a fraction of the power and performance found in a Single Frequency detector (illustration A).

Simply stated, detectors represented by illustrations (A) and (B) contain the same amount of power and potential performance. The difference is that a Single Frequency detector focuses 100% of its power at a nominal frequency that finds the most treasure over the most common ground conditions. In contrast, a Multiple Frequency detector divides its power into 2, 18 or more frequencies for the purpose of overcoming heavily mineralised soils. So, when deciding which technology—Single or Multiple Frequency—is best for you, consider where you will do most of your metal detecting. If you will hunt in difficult conditions where mineralisation is a problem, use a Multiple Frequency detector. However, if you plan on hunting in lightly mineralised fields, use a Single Frequency detector.

be mathematically represented by the combination of many different frequency components, known as the Fourier Transform *(see illustration on page 25)*.

In much the same way, the magnetic spectrum of the Pulse Induction wave contains multiple frequencies or harmonics. Stating that a detector operates with 18 frequencies, for example, only shows its pulse wave is made up of 18 significant harmonics. It does not mean a detector is 18 times better, nor the equivalent of 18 detectors operating at once *(see illustration above)*.

Pulse Induction/Multiple Frequency is simply another way of driving the detector's magnetic energy into the ground. However, just as sonar may be superior to other sensing technologies in certain applications, there are some environments where Pulse Induction has an advantage over Single Frequency technology and vice-versa.

Multiple Frequency technology, which is most commonly implemented as Pulse Induction, was initially designed specifically for use in highly mineralised soils like those found on the beach or in the gold fields. Wreck divers and prospectors who hunt in areas of high ground mineralisation generally use specialized PI detectors due to the extra stability they offer. However, target ID accuracy will typically be poor with these detectors, primarily in the region of iron identification. Many PI detectors, in fact, do not include a target ID at all. In addition, other important features found on Single Frequency detectors, such as True Size and Depth measurement, Tone ID and advanced discrimination modes either suffer or are not available on Multiple Frequency instruments. Such PI detectors are also less energy efficient, often requiring large and heavy battery packs.

Searchcoil Essentials

The second essential component of metal detecting is the searchcoil—also known as search head, loop, coil or head.

Choosing the right searchcoil from all the different *sizes, shapes, configurations* and *construction* types available is one of the most important aspects of becoming a successful treasure hunter. Searchcoils are generally plastic, molded resin or epoxy-filled housings that contain many thin copper wire windings that send and receive the signals. Most quality manufacturers make watertight searchcoils that can be totally immersed in water without affecting any of the wire loops inside.

The searchcoil attaches to the end of the detector's stem. A coil cable usually winds around the stem and plugs into the detector's control housing. Most searchcoils operate with two separate internal sets of coiled wires, a transmit coil (TX) and a receive coil (RX). Mono coils can be different since one coil acts as both the TX and the RX. When the metal detector is turned on, an electromagnetic field generated by the transmitter winding flows out into whatever medium is present—soil, rock, water, sand, wood, air or whatever else might be encountered. When a metallic object is within this generated magnetic field, it will create a distortion in the magnetic field. The RX coil will sense this distortion and send a signal to the control housing. A searchcoil's detection pattern is determined by the combination of the TX's generated field pattern and the RX's sensing field pattern.

Because searchcoil construction quality can vary, it is important to choose a reputable manufacturer. Consult fellow detectorists, read reviews and ask the advice of your dealer before opting for the cheapest coil on the market. Although some inexpensive metal detectors may come with a permanently mounted searchcoil, most quality detectors have a stem assembly that allows the user to switch searchcoils based on different hunting needs.

Searchcoil Size: As you gain experience in the field, you will learn that there is no such thing as "the best searchcoil" for all applications. The same coil that performs well in a ploughed field may be challenged on a trash-filled beach. Searchcoils are thus available in a wide range of sizes.

Bigger searchcoils do not necessarily guarantee greater detection depth with all targets. Larger targets will certainly be detected at greater depths with larger searchcoils. Small-sized objects (such as a Euro) can be missed when hunting with a searchcoil that is too large. *(See illustration on next page.)*

A metal detector typically puts out a fixed amount of energy regardless of the size of the searchcoil. However, the volume or region into which that energy is transmitted is determined by the size of the coil. A larger coil transmits into a larger region, both deeper and wider while a small coil transmits into a smaller region. Therefore, the energy will be much more concentrated and intense beneath a smaller coil. Conversely, the bigger coil takes that same amount of energy and spreads it into a larger and deeper area, but with less concentration and intensity. As a result, the smaller coil with its concentrated energy field is able to detect small objects better than the larger coil with its dispersed field. It is true that

WHAT SIZE SEARCHCOIL SHOULD I USE?

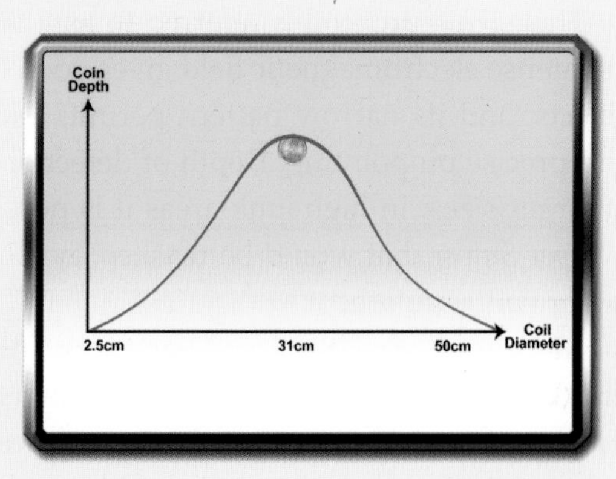

Large searchcoils offer the ability to cover more ground and search deeper. There is, however, a relationship between searchcoil diameter and effective search depth. As a rule of thumb, use a standard size coin (such as a Euro) to gauge effective detection depth.

For example, a 10-cm coil will very effectively locate a Euro at 10 cm depth. A 31-cm coil will effectively find the same Euro at about a 31 cm depth. As seen in the illustration above, however, the coin-sized target becomes more difficult for searchcoils to detect effectively as the coil diameter continues to increase. By the time you reach a 50-cm coil, the coin-sized object can only be detected at perhaps six centimeters depth. It is important to remember that the magnetic energy of the searchcoil is much more heavily concentrated in smaller diameter coils.

bigger searchcoils will search deeper but the size-versus-energy concentration has an effect on what size target can be discovered as the size of the coil increases.

For effective hunting, you should use smaller coils for smaller, shallow targets. Use larger coils to detect larger, deeply buried objects. For general-purpose hunting, use 18 cm to 25 cm coils.

Small Coils (10–15 cm Diameter): Because the magnetic field of a small searchcoil is concentrated within a small area, that coil size is the best choice for hunting where a lot of metal debris is encountered. This allows the user to maneuver through and around trash to locate good targets, especially when searching in tight places where large searchcoils cannot go. However, because a small searchcoil provides less coverage and less depth per sweep, more scans will be required to cover a search area. It only takes a tiny piece of metal to disturb the magnetic field below the searchcoil when using smaller searchcoils with more heavily-concentrated fields. As long as the object is within reach of the diameter of your coil, it will be more easily detected by these smaller coils.

This size searchcoil is referred to as a *Super Sniper* by Garrett. Its intense electromagnetic field gives good detection of very small objects, and its narrow pattern permits excellent target isolation and precise pinpointing. Depth of detection is not as great as that of larger sizes. In high junk areas it is possible to find coins with a *Super Sniper* that would be masked by adjacent junk signals if a larger coil were used.

Medium Coils (18–25 cm Diameter): This size searchcoil is furnished with most detectors because it is usually the best size for coin hunting and other general purpose uses. These searchcoils are lightweight, have good scanning width and are sensitive to a wide range of targets. Small objects can be detected, and good ground coverage can be obtained. Scanning width is approximately equal to the diameter of the searchcoil. Depth of detection is excellent for most targets with a searchcoil of this size.

Large Coils (30 cm and Larger Diameter): Searchcoils this large, while able to detect larger coins and artefacts at greater depths, are also classified as the best searchcoils for cache and artefact hunting. Precise pinpointing is obviously more difficult with the larger sizes, and their increased weight usually necessitates the use of an arm cuff or a hip-mounted control housing, especially when the detector is used for long periods of time. You will want to use this larger searchco"il when you expect to search deeply for larger coins or any other treasures.

When you are hunting coins, how do you know when to switch from your "standard" size to the larger searchcoil? The search conditions often help dictate when you should make a switch. For example, suppose you locate a target at the fringe of detection. You know from the weak audio signals that you are at the outer edges of your detector's capability with this searchcoil.

By using a larger size, you will generally detect deeper but run the risk the losing smaller targets as the coil size increases. Larger coils also present the increased risk that you will pass over iron

SEARCHCOIL DETECTION DEPTH

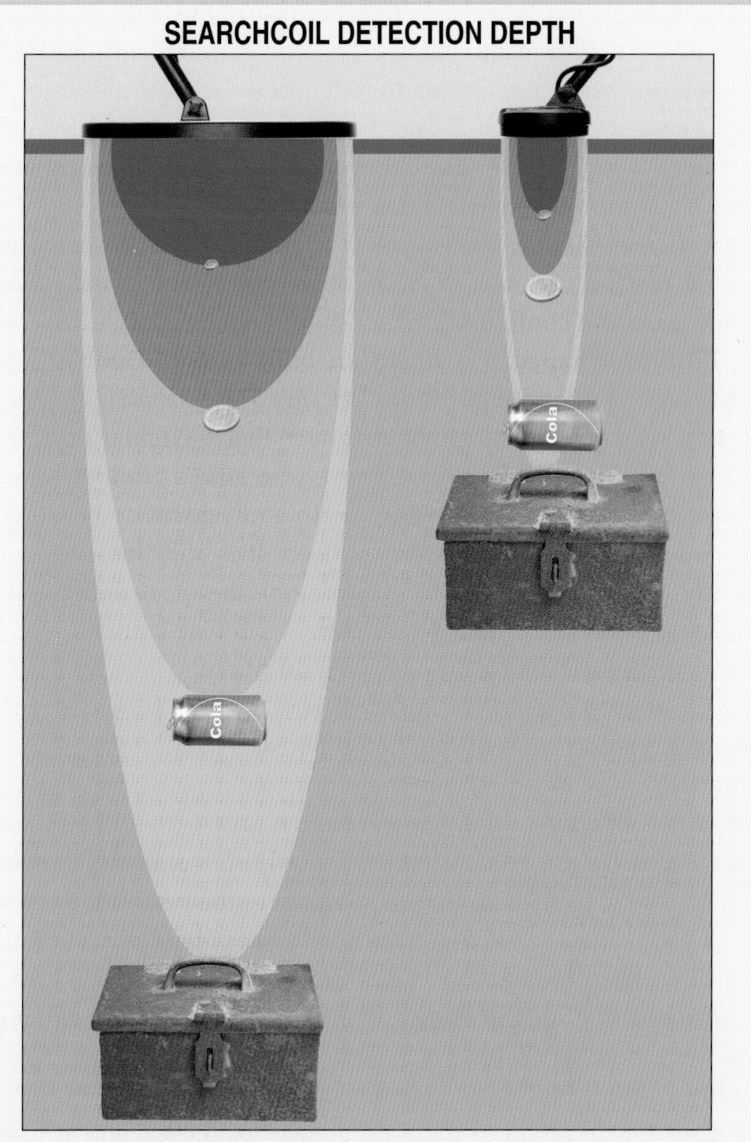

These illustrations compare the average detection capabilities of a larger diameter searchcoil (31 to 38 cm) versus a small searchcoil (12 cm diameter). Larger coils achieve greater detection depths.

Larger items can be detected at greater depths than smaller items. Each shaded region represents the detection limits for the particular sized item shown. The items shown are a tiny silver coin, a 2-Euro coin, a soda can and a large money box.

rubbish and coins at the same time because of the increased region into which they transmit energy. The iron garbage can "mask" the good target you were hoping to find; more on target masking will be discussed in Chapter 3. The "Searchcoil Detection Depth" illustration *(see page 31)* depicts the average detection depth capabilities of a large coil versus a much smaller coil.

Searchcoil Shape: The two most common shapes of searchcoils are *circular (round)* and *elliptical*. Searchcoil shape is *independent* from searchcoil configuration and should not be confused.

Elliptical shaped searchcoils are built for both Double-D and concentric search configurations *(see page 37 for comparison)*. An elliptical searchcoil is more maneuverable than a circular searchcoil due to its elongated length. However, because a **circular searchcoil** has slightly more detection depth and sensitivity because of the maximum loop area being utilized, it is still the most commonly used shape. Circular coils provide the most uniform pinpointing because of their symmetry. A circular coil can also be either concentric or Double-D in configuration.

An elliptical-shaped coil's narrow width provides more maneuverability for working in tight areas. An elliptical 15x25 cm coil is 15 cm wide and 25 cm long from heel to toe versus a 25 cm circular coil, which offers greater search area with 25 cm in each direction. The equivalent diameter of a 15x25 cm coil is the average of the two dimensions, or equivalent to the performance of an 20 cm circular coil. Again, the elliptical coil potentially gives up a little bit of depth but offers better maneuverability in tight areas.

Searchcoil Configurations: There are a variety of searchcoil configurations available, each with various benefits for different hunting applications and ground conditions. The "configuration" of a searchcoil refers to the arrangement of the TX and RX coils within the searchcoil shell. There are basically five configurations: Concentric, Mono, Imaging, Double-D and 2-box. The searchcoil's *configuration* is more important than the shape of the coil.

The **concentric configuration** consists of a TX coil and RX coil which are usually circular and arranged *(see adjacent page)*. The

SEARCHCOIL CONFIGURATIONS AND SHAPES

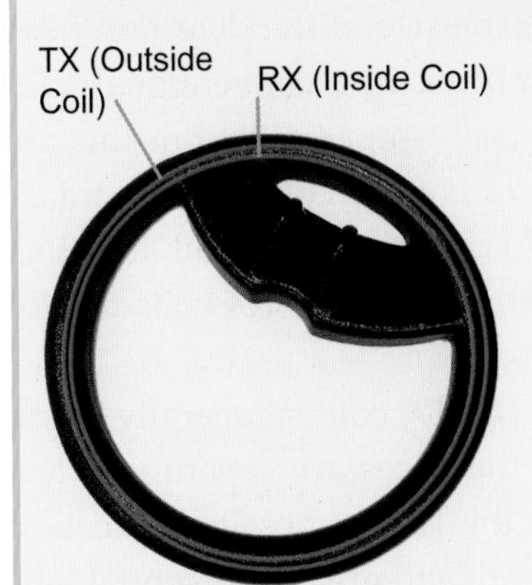

Mono Searchcoil with Circular Shape

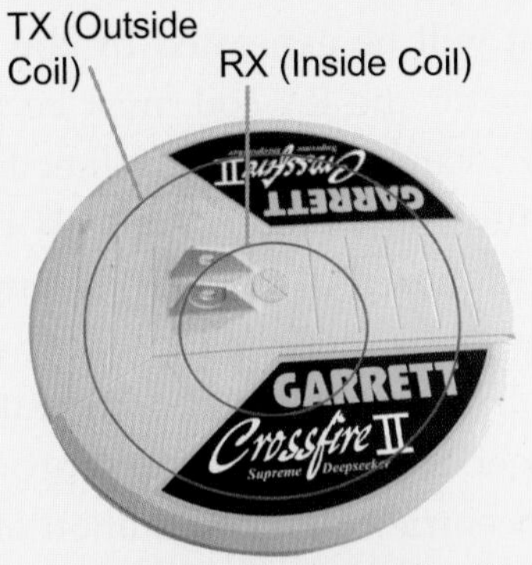

Concentric Searchcoil with Circular Shape

Double-D Searchcoil with Eliptical Shape

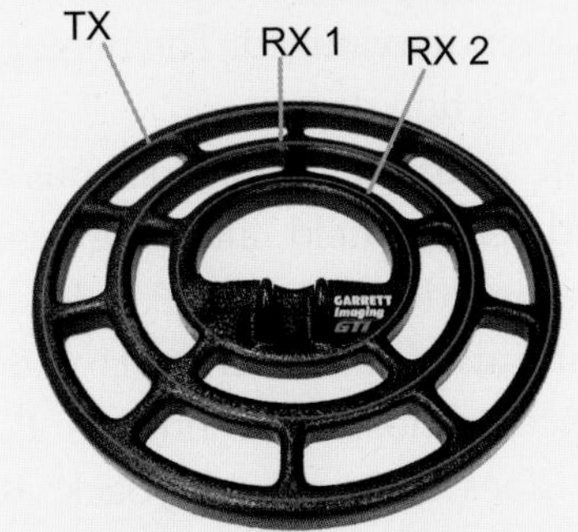

Circular, Imaging Searchcoil

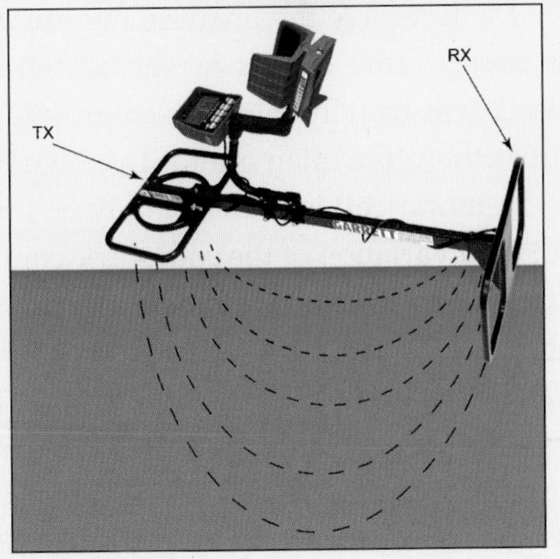

Depth Multiplier or 2-Box Configuration (for larger, deep targets)

advantage of this configuration is that both the TX and RX coils can be wound as large as possible within a given searchcoil diameter. The size of these windings dictates the size of the detection field that will be generated by the searchcoil. In a conventional VLF detector the RX coil must be physically separated from the TX. They cannot be stacked directly on top of each other or the transmitter coil would completely overload the receiver coil and thus prevent detection. (For PI machines, the TX and RX coils do not have to be separated.)

In the concentric configuration, the TX coil is generally made larger than the RX coil to provide the necessary separation. The concentric coil configuration offers the largest possible detection field, greatest potential detection depth and greatest potential sensitivity because the coil loop area within the searchcoil's housing has been maximized. The bigger the coil windings are, the greater the detection depth.

In addition, concentric coils also provide the most symmetrical detection field, allowing ease in pinpointing and consistency in target identification. For these reasons, they are the most commonly used and such coils will provide the best overall performance in most environments. For all of its virtues the concentric coil does have its drawback. Because it transmits the greatest amount of energy into the ground, it is also the coil configuration that will receive the greatest amount of interference from ground minerals. This results in substantial loss of performance when used over heavily mineralised ground. Detectorists working such areas therefore often opt for Double-D searchcoil configurations.

A **mono-coil** is available only on Pulse Induction (PI) detectors and is a variation of the concentric configuration. In VLF detectors, the TX and RX coils cannot be located on top of each other because they are transmitting constantly. In a PI machine, the mono-coil can be manufactured with the TX and RX coils located together or as a single coil acting as both TX and RX. The detection and performance characteristics of the mono-coil are essentially the same as the concentric in that it provides the maximum possible sensi-

Searchcoils are available in many shapes and sizes for various hunting uses. Top row *(left to right)*: 6.5" x 9" elliptical concentric coil, 12.5" circular imaging coil, and 10" x 14" elliptical Double-D coil. Bottom row *(left to right)*: 4.5" *SuperSniper* circular coil, 9.5" circular imaging coil, and a 9" x 12" elliptical concentric coil.

The mono-coil for Pulse Induction detectors is lightweight and offers better detection capabilities in most soil conditions.

tivity, but suffers some performance loss in mineralised ground. The mono-coils on PI machines offer better sensitivity and depth because of their maximum loop area. In the most extreme mineralisation conditions, a Double-D searchcoil will offer performance advantages.

An **Imaging searchcoil** is an enhanced version of the concentric configuration that features an additional RX coil. This extra coil provides the detector with additional target information necessary for true target-depth perception and true target-sizing capabilities. With this additional sizing information, the detector can more fully characterize a target and for the first time distinguish between trash and good targets of the same conductivity (e.g. a coin vs. a soda can). Garrett's *GTI (Graphic Target Imaging)* metal detectors have imaging searchcoils that use this advanced technology.

Imaging searchcoils use a second RX coil that acts like a person's second eye. The second receiver is smaller than the first receiver and helps judge depth and size of objects by offering a second perspective.

The **Double-D coil configuration** is designed to significantly reduce ground interference and, thereby, recover the performance lost by a concentric coil over mineralised soil. This configuration is called DD because both TX and RX coils are in the shape of a "D." The interior DD coils can be housed within a searchcoil that is elliptical, round or even square in shape.

The Double-D arrangement of TX and RX coils produces a canceling effect of ground signals. The positive detection field of the DD is located beneath the overlapping center section from front-to-back. The remaining portion of the coil actually produces negative (i.e. canceling) detection fields. It is this canceling field that allows the DD coil to cancel the effects of mineralised ground and maintain its performance over such ground *(see illustration on page 37)*. Double-D coils are useful when hunting highly mineralised grounds and also experience less saltwater interference.

A 25 cm DD searchcoil containing slightly overlapping 15 cm RX and TX coils does not offer the same 25 cm search area that a

DOUBLE-D VERSUS CONCENTRIC SEARCHCOILS

The DD searchcoil has a much different detection field than a concentric coil. The DD's positive detection field is highly concentrated in the center, with negative fields on either side which provide the cancellation of the ground signals.

The concentric searchcoil provides the largest possible detection field and greater detection depth in normal soil conditions but is more susceptible to ground minerals. The concentric coil also offers greater ease in pinpointing targets and in target ID.

As depicted, the detection depth with a DD coil becomes superior to that of a concentric coil as the level of ground mineralization increases.

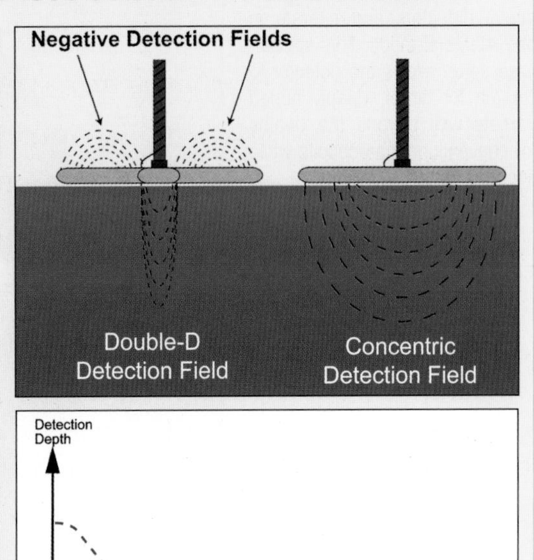

25 cm concentric coil offers. Because of its small positive detection field, the DD is inherently less sensitive than a concentric searchcoil of the same size over non-mineralised ground. The Double-D will, however, outperform the concentric coil over mineralised ground. For this reason, it is highly recommended for hunting mineralised ground often found when prospecting and artefact hunting.

Another type of searchcoil configuration is the **depth multiplier or 2-box configuration** for ultra-deep detecting. In this configuration, the TX and RX coils are physically separated by a significant distance (e.g. 1 meter). This provides a lightweight, manageable means of achieving the performance of a 1-meter-diameter searchcoil that can be physically managed. Because of its large size, and

Depth multipliers or 2-box searchcoil configurations mount to select metal detectors and operate in the All-Metal mode. This specialized setup allows the detector to search for larger, deeply buried targets well beyond the depths of conventional searchcoils while ignoring small trash items.

consequently large detection field, the 2-box is the best choice for detecting large, deeply buried objects such as hoards. Also, because of its large detection field, it ignores objects smaller than about 7 cm in diameter. This characteristic is advantageous when hunting in areas heavily littered with small trash objects. The depth multiplier configuration is recommended when searching for coin or weapons caches, large artefacts, safes or cannons.

Garrett offers an exclusive, enhanced version of this 2-box configuration known as the *Treasure Hound with Eagle-Eye* pinpointing. This version incorporates an additional pinpointing coil in the front for precise target location.

Searchcoil construction: There are three main types of searchcoil construction. An **air-filled** searchcoil is basically a plastic shell that houses the RX and TX coil windings. This is the lightest weight configuration but it is also the least durable without a solid internal structure for support. Air-filled searchcoils traditionally do not seal well and can suffer from leakage in wet environments which can quickly ruin a searchcoil.

The second type of coil construction is a plastic shell which is **foam-filled**. This can be pre-cut foam or expandable foam that fills the housing. This provides a relatively light-weight coil with some added structure for the internal coil windings. This coil still suffers from water leaks because it is not truly watertight.

The third type of searchcoil construction is an **epoxy-filled** searchcoil. It is the most durable version as a rock-hard epoxy protects the windings. The only drawback is that this makes the searchcoil heavier; therefore, most epoxy-filled coils have openings cut through the coil versus being a solid housing. An advantage of these openings is visibility through the coil to see the ground as a target is pinpointed. The epoxy-filled coil is rugged and offers the best protection from water because the internal windings are fully sealed.

Make certain you understand the water-resistant capabilities of your searchcoil. *Splashproof* indicates that operation will not be affected if a small amount of water gets on the searchcoil, such as light rain or moisture from wet grass. *Weatherproof* (sometimes called waterproof) means that the searchcoil can be operated in heavy rain with no danger from the moisture. *Submersible* indicates that a searchcoil can be fully submerged as deep as the cable connector, without affecting the detector's operation.

Avoid "bargain" searchcoils. A good searchcoil is vital to the success of a metal detector. It should be light, but it must also be sturdy and capable of rugged treatment. A good searchcoil has a tough exterior that will not abrade or crack easily on rough ground. Searchcoils must withstand the greatest abuse of any detector component because they are constantly being slid across the ground, bumped into rocks and trees, submerged in water and generally mistreated in every way.

Skid plates or coil covers are available that protect the surface of a coil from being damaged or heavily scratched by rough ground, rocks, roots and other obstacles that a searchcoil brushes against. Coil covers prevent the search head from eventually becoming so scuffed that cracks form which can cause leakage into the inner coil wires, rendering the coil useless. Such scuff covers are thus a wise investment to prolong the life of valuable search-coils.

All-Metal (Non-Motion) Detectors versus Discrimination/ Target ID (Motion) Detectors

The earliest metal detectors operated in an All-Metal mode without any discrimination. They simply identified all types of metal that their searchcoil passed over. Many European detectorists today wish to ignore small iron and pull tabs, which requires a detector with discrimination settings that can be applied. It is important to note some of the key differences between the two primary detector types that are sold today.

In a true **All-Metal**, non-motion detector, your machine is detecting all types of metals, responding to all metals it encounters. One of the advantages of an All-Metal detector is that because it is not trying to analyze whether a target is worthy or not, the detector can provide the greatest possible depth and sensitivity—often achieving depths up to twice that of a discrimination detector. The other advantage of All-Metal detectors is that operate without requiring the searchcoil to be in motion. This static, or very near static, operation mode is the reason that All-Metal machines are referred to as *Non-Motion* or Motionless detectors.

Such a detector allows you to approach a target, hover over a target or creep around a target with your searchcoil. You will hear every nuance and every subtle increase or decrease in target response as you approach it and leave it. When veteran detectorists hear every characteristic of their target, they can literally hear the shapeor length of a target as the audio changes. This continuous audio response helps distinguish a target's features.

The principle drawback to this continuous audio response is that you also hear every subtle nuance of the environment. The biggest environmental factor, of course, is the ground itself, which contains all types of minerals. This can include ferrous minerals, conductive minerals and moisture which all produce responses. Some rocks—known as "hot rocks"—contain concentrated minerals and can produce a significant audio response. You thus have to become adept at distinguishing the sounds of ground response mixed with the sound of a good target's response.

The first thing you must thus do with an All-Metal detector is ground balance it. Without ground balancing, the ground itself will produce such strong signals that you can not hope to find a target. All you hear is *ground* when you move the coil until you ground balance your machine. This is literally adjusting and tuning the electronics of your detector so that the ground response is essentially ignored. As soon as the ground mineralisation shifts, however, you must ground balance your detector again to this new environment. This can be challenging for new detectorists.

Circuitry or signals in an All-Metal detector can potentially drift, adding another challenge to the operator. Drift in the ambient background signal can be caused by: drift in the electronics (although modern electronics are extremely stable); temperature drift as you might move from the hot sun into a cool, shady spot; and changes in the ground mineralisation. You must remain aware of this ever-changing environment because of the sensitivity of an All-Metal detector.

A good quality All-Metal mode detector has features to help overcome the difficulties presented by ground balancing and drift. To keep these effects to a minimum, a quality All-Metal detector has features like Auto Threshold and Auto GroundTrack. With Auto Threshold, the detector automatically monitors the ambient signal; if it starts drifting because of temperature changes, etc., the Auto Threshold will work to suppress this. In Auto GroundTrack, the detector continually monitors the ground conditions and adjusts itself to these changes.

Discrimination detectors with pinpointing abilities are basically switching into an All-Metal mode to allow you to continually hear the target response as you move the searchcoil over the target. Examples of this are the *ACE 250* and *GTP 1350* detectors by Garrett, which temporarily switch into All-Metal operation when you hold down the Pinpoint button.

When the *ACE 250* and *GTP 1350* in their Pinpoint modes put the machine into an All-Metal mode, it is similar to comparing an amateur sports team to professional teams. You get the advantages of continuous audio feedback, and you have static operation for hovering over targets without losing the signal; what you don't have in this Pinpoint mode is all the other support circuitry that a good All-Metal mode has such as the automatic threshold and ground tracking functions. Because this pinpointing function is thus not as stable as that of an All-Metal mode detector, opting to hunt while holding down the Pinpoint button will not be effective. You will have to constantly release and reset the Pinpoint button because of the changing environment. It is quite effective, however, during the short time normally required for pinpointing.

The other type of detector is the **Discrimination** or **Motion** detector which does not give continuous audio feedback. The detector effectively hunts silently in Discrimination mode until it identifies and signals a target. One of the advantages of the Discrimination mode is that it frees the operator from all the distractions that comes with the All-Metal Mode by not responding to everything it encounters. The detector pre-screens everything it analyzes before announcing an item it interprets to be a good target. The user is able to pick and choose which targets should respond and which targets should not respond, such as iron.

In order for this type of detector to operate correctly, the searchcoil must be in motion because discrimination detectors introduce filtering to minimize the ground signal. This filter must be in motion to be effective which requires the coil to be moving as it discriminates or identifies target metals. When you hold the searchcoil perfectly still above a target with a discrimination detector, it loses the signal.

A Discrimination (Motion) detector operated in a "Zero" or "All-Metal" mode is essentially detecting all metals, but this is misleading because of the nature of the discrimination filters. When you are using a Discrimination mode of any kind you can never experience a true All-Metal mode because the motion detector always maintains some amount of filtering to eliminate the response of ground and other small perturbations. A true All-Metal detector on the other hand has no discrimination filters, allowing it to signal all metals. On most Discrimination detectors, iron may be the lowest conductivity target identified on the target ID scale, but even lower on a scale than iron are the electrical and magnetic properties of the ground.

Only a true All-Metal detector (i.e. no discrimination filters) can respond to the entire range of electrical and magnetic properties. *A Discrimination detector with an "All-Metal" mode is not a "true" All-Metal detector;* this mode should more properly be termed "Zero" discrimination as there is always some discrimination occurring in a motion detector.

You can determine whether your detector is a motion or non-motion type with a simple test. Sweep a coin or other target under the searchcoil to produce a response. When you hold the target perfectly still under the searchcoil, a motion detector will cease to respond. This type of machine requires the searchcoil to be in at least a slight amount of motion to detect a metallic target.

Target Identification with Motion Detectors

A detector's target ID refers to its ability to identify a target. *The conductivity, permeability, thickness, size, shape, and orientation of the target all play roles in determining where it will read on its target ID scale.*

Every metal has a certain electrical characteristic and a certain magnetic characteristic. The electrical characteristic is *conductivity*. The magnetic characteristic is called *permeability;* a magnet will attract such metals. These two characteristics dictate how metal will distort the magnetic field transmitted by the searchcoil.

The most highly conductive materials, such as pure silver, have no magnetic properties at all and are known as purely conductive *non-ferrous* metals. When the magnetic field from a detector's searchcoil bombards a purely conductive metal, eddy currents form on the surface and produce a secondary field that reflects back. In other words, that piece of metal acts like a mirror, reflecting what has been transmitted against it. What is received back is basically a mirror image of the field that was transmitted *(see illustration on page 45)*.

An extremely thin conductive metal will allow some of the magnetic field to pass through it; the thicker the conductive metal, the more it will perfectly bounce the magnetic waves back. Thick, highly conductive metals which most strongly reflect the magnetic waves back to the searchcoil will register highest on the target ID scale. Pure silver is the most conductive metal of the treasures you will usually encounter.

The other extreme example is a metallic object that has very low conductivity but is very magnetic or *ferrous* (i.e. iron). Rather than trying to reflect the magnetic field, a magnetic item funnels the field through itself and out the other end *(see illustration on page 45)*. An iron bolt in the magnetic field will draw in the magnetic fields from the coil and create a distortion the exact opposite of the purely conductive silver. These ferrous (magnetic) metals show lowest on the target ID scale.

What does all this mean? Again, electrical conductivity and permeability (magnetic characteristics) are the two major factors that determine how the object will distort the magnetic field (i.e. conductive metals will reflect the field and magnetic metals will draw the field inward). The third mechanism that occurs is *energy dissipation* or energy loss. This happens when eddy currents on the target's surface interact with the metal's resistance and actually begin dissipating energy in the form of heat. The higher the metal's resistance (i.e. the lower the conductivity) the greater the energy dissipation. Very little energy is lost in highly conductive metals such as silver. All metals have a certain amount of energy loss

TARGET SIGNALS: Conductivity versus Permeability

This top illustration depicts the magnetic field generated by a metal detector's searchcoil.

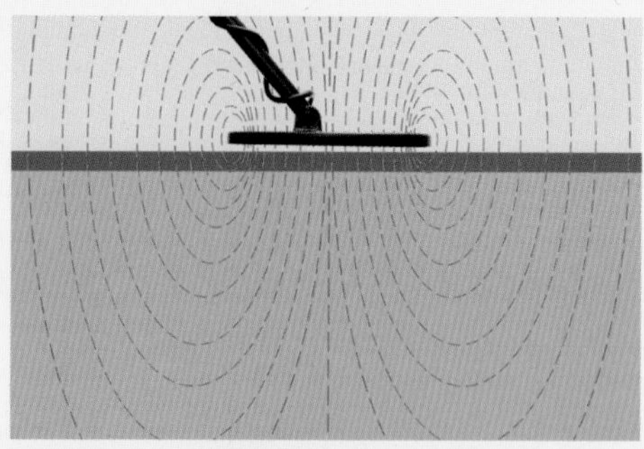

Conductive Target

This center illustration depicts the distortion of the magnetic fields when it encounters a highly conductive target such as a silver coin. This purely conductive target produces currents that reflect the field almost like a mirror.

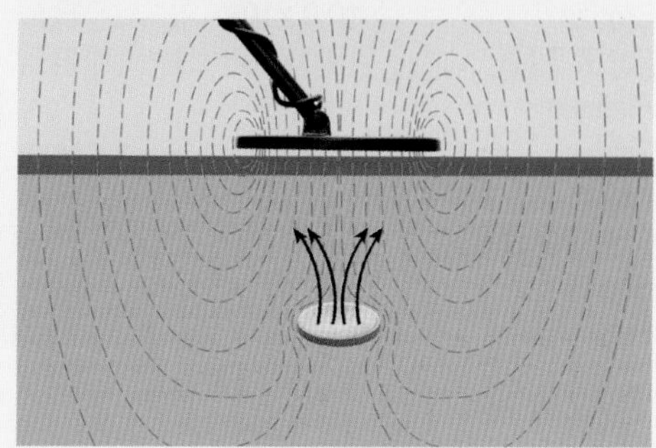

Magnetic Target

This lower illustration depicts the distortion of the magnetic field when it encounters a magnetic (ferrous) metal item. This object pulls in the field (versus bouncing it back) through itself, creating a response that is the exact opposite of the highly conductive coin.

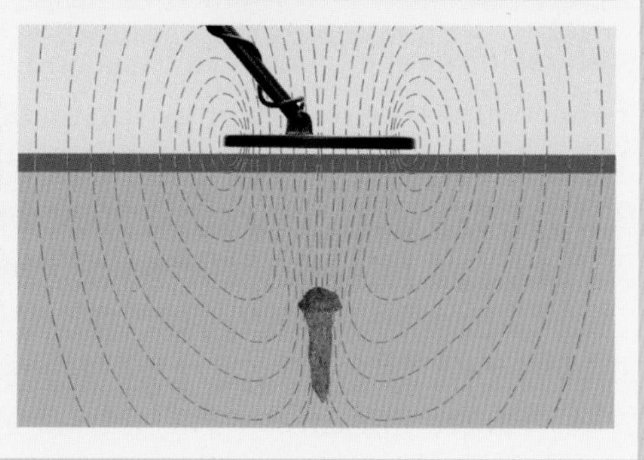

associated with them. *A metal object's target ID is determined by a ratio of energy dissipation to field distortion.*

The levels of electrical conductivity of various metals are shown on displays or meters on metal detectors. Scientists measure the levels of conductivity in Siemens per meter (S/m) with an ohmmeter by measuring how easily electrons move through a metal object. Iron is on the lower end of the conductivity scale along

ELECTRICAL AND MAGNETIC PROPERTIES OF COMMON MATERIALS

Material:	Conductivity σ (S/m)	Conductivity Ranking	Magnetic Permeability μ_r
Silver	62.9 x10^6	Excellent	1
Copper	59.8 x10^6	Excellent	1
Gold	41.0 x10^6	Good	1
Aluminum	35.4 x10^6	Good	1
Zinc	16.9 x10^6	Fair	1
Brass	15.7 x10^6	Fair	1
Nickel	14.6 x10^6	Fair	250
Bronze	10.0 x10^6	Fair	1
Iron & Steel	10.0 x10^6	Fair	400 (typical)
Platinum	9.5 x10^6	Poor	1
Tin	9.1 x10^6	Poor	1
Lead	4.8 x10^6	Poor	1
Stainless Steel *	1.4 x10^6	Poor	1 (typical)
Sea Water	4.0	Very Poor	1
Fresh Water	0.001	Extremely Poor	1

* Non-Magnetic

with stainless steel. The electrical and magnetic characteristics of metal, coupled with its energy dissipation are again key factors in determining how a metal object registers on the detector's target ID scale. (See chart above to see how conductivity and permeability both come into play to help rank items.)

In addition to the three scenarios described above, target ID is also determined by several other factors:

- **Target thickness**—The thickness of a metal object plays a huge role in its target ID reading, far beyond any minor effect due to the alloying of metals. The thicker a conductive object, the greater its ability to conduct larger eddy currents that reflect more of the magnetic field. The material's inherent conductivity and its thickness together are the primary factors determining what the "effective conductivity" or effective target ID will be. Even though large pieces of silver register highest on the scale, an equally pure piece of silver hammered paper thin will not register the same.

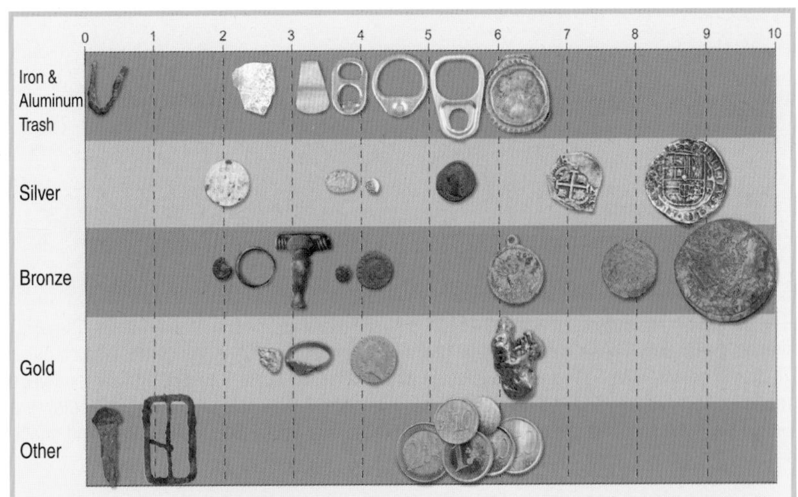

This target ID scale represents where various European targets and metallic trash will register on most metal detectors. The conductivity scale, running from low conductive metal items such as iron on the far left to highly conductive objects on the right, is displayed in a variety of meters, graphic interfaces and numerical systems on today's different detectors. (It should be noted that some detectors, such as Garrett's *GTI* series, use a 12-scale graphic target identification system.)

As illustrated above, metal composition of a target (such as silver) is only one variable that determines where a target will register on a scale. The target's size, thickness and orientation in the ground all affect how a target will register.

See Chapter 3 for more details on target discrimination.

- **Target Shape and Orientation**—For ferrous objects (e.g. iron and steel), the shape and orientation of the target can have a significant effect on the target ID. For example, a thin steel washer or flattened bottle cap lying flat in the ground can produce a high target ID reading, similar to a coin. This is because the object's horizontal orientation is presenting a large surface area to the searchcoil, thereby allowing eddy currents to flow on that large surface and reflect the magnetic fields. This same steel washer or flattened bottle cap when resting vertically on edge is no longer presenting a large surface area. Rather, it is presenting a long, narrow path which channels the magnetic field inward through the object (same as the iron spike shown in the illustration on page 45). As a result, the steel washer or flattened bottle cap oriented verti-

The shape and thickness of a metal object is much more important in determining Target ID than its metal alloy. These 10k gold items (actually composed of 75% silver content) found by an *ACE 250* detectorist showed on the left side of his Target ID scale. The 1865 medallion (since made into an earring) showed on the *ACE* scale under foil. The gold ring at right hit below the 5¢ on his detector. Their target reading was more heavily influenced by their size and shape versus their gold to silver alloy ratio.

cally will produce a low target ID reading of iron. This same effect can occur for most any piece of iron—such as a piece of broken plough blade—which, when flat can produce a high target ID, and when vertically oriented will produce a low target ID of iron.

In summary, Discrimination detectors with target ID capabilities allow you to focus on the metals you want to detect and ignore the metals you don't want to detect. This time-saving capability revolutionized the metal detector industry when it allowed hobbyists to pick and choose targets. The drawback to Motion detectors with discrimination is that the detector must continually analyze all targets that are encountered. The signal that measures magnetic field distortion is extremely susceptible to mineralised ground, making Discrimination detectors more challenged by adverse mineral conditions (and wetted salt sand) than All-Metal (Non-Motion) detectors.

The magnetic properties inherent in all soils factor into the true reporting of target ID. In even the most inert (pure) ground conditions, the ground signal skews the accuracy of the target ID to a certain degree. Of course, target ID is extremely accurate in air tests without the presence of ground mineralisation. In inert ground, target ID is again very accurate but steadily declines as the amount of mineralisation increases. White, dry beach sand, with no magnetic black sand or other conductive minerals, is also an excellent area for accurate target identification.

Many detector target ID scales report on a 10- or 100-point scale. Some high-end Garrett models also report on a 24-point scale. They are all reporting the same physics concerning the ratio of field distortion versus energy absorption. The 100-point scale systems claim to offer better distinction between target readings than a 10- or 24-point scale.

The reality is, however, that ground minerals and other environmental factors will produce the same percentage of fluctuations in the target ID reading, regardless of the number of points in the target ID scale. For example, if the ground is creating a ± 10% fluctuation in the target ID, then a target which reads 4 to 6 on a 10-point scale will likely read 40 to 60 on a 100-point scale. In this situation, the 100-point scale is not providing any more information than the 10-point scale. The extra resolution of the 100-point scale is only beneficial in those rare ideal circumstances where external fluctuations can be minimized.

Signal Processing

The fourth essential component of a metal detector is the way in which it analyzes received signals to determine target information. All metal detectors, regardless of their technology, transmit an analog signal through space and ground and must determine how it will interact with a metal target.

Once those signals return to the detector through the searchcoil, the technology of the machine's signal processing comes into play. Older units and even some less sophisticated detectors still on the market perform their signal processing and analysis via analog circuitry. By the mid-1980s, *microprocessors* were being incorporated into metal detectors. This immediately eliminated the variability and drift inherent with analog components and filters. Microprocessors use digital processing, which is more absolute, thus offering greater consistency and reliability. The introduction of microprocessors also permitted simpler adjustability. Instead of several knobs to adjust, the microprocessor often permitted the use of touch pads and LCDs that offered greater adjustability, repeatability and flexibility without the variations of analog circuitry.

The next jump forward was *digital signal processing (DSP)*, comparable to a microprocessor on steroids. This powerful, high-speed math machine took even more of the filtering and signal processing away from the analog circuitry to a digital system that is more consistent and accurate. Detectors that utilize digital signal processing are generally more high-end, more expensive and among the most sophisticated detectors on the market. Don't be misled by brands with microprocessors that simply process signals digitally (and claim to be DSPs) versus a true digital-signal-processor instrument. Analog circuitry and microprocessors alone simply cannot process the same amount of signals as fast and as reliably as a metal detector with DSP.

Description of Metal Detector Features

Your decision in purchasing a detector should be driven to some degree by the particular features you desire. Some metal detectors will allow you to make adjustments with knobs or dials while others feature push-buttons.

In any event, the following list will describe some of the key features you should consider before making a purchase. Some of these you might find to be "desired"; others you may considered to be "required":

• *Discrimination Mode*—This indicates a detector's ability to decipher between treasure targets and trash targets. This is usually a switch, dial or push-button selection that allows the user to "tell" the detector what items to "accept" and what items to "reject." This allows hunting in trashy areas without detecting and digging every piece of rubbish. For this reason, a detector with discrimination modes is considered a required feature by most detectorists. A disadvantage of any detector employing a Discrimination Mode is that it achieves less depth than one with an All-Metal Mode.

• *All Metal Mode*—A true All-Metal mode detector provides the greatest possible depth and sensitivity. It also provides continuous audio response of the target to discern subtle characteristics of the target such as size, shape, etc. An All-Metal mode machine can be at a disadvantage—or virtually useless—in very trashy areas due to excessive target responses. *(Refer back to the All-Metal versus Discrimination section in this chapter for more details.)*

• *Notch Accept/Reject Discrimination*—This feature on many detectors enables groups of metals to be accepted or rejected by notching them in or notching them out. In Chapter 3, the dangers of using notch discrimination to remove unwanted "trash" targets such as pull tabs will be illustrated. Even with the elimination of ring pulls, you will also be eliminating any gold rings which have the same conductivity as these ring pulls. Many detectorists will opt to eliminate some of the iron targets on the left side of the scale, again knowing that they are also potentially missing some good artefact targets.

Notch discrimination is particularly useful if you are searching for a specific type of target and wish to eliminate all other metals—thus increasing your ability to find the desired target metal. Rallyists who are searching for specific prize tokens can notch out other unwanted targets in order to speed their search for such tokens. *(See discrimination notches and other detector features on the control panel image on page 52.)*

• *Automatic Ground Track*—This feature may also be referred to as "automatic ground balancing." Such detectors more or less maintain optimum ground balance against mineralisation in order to prevent the need for continuous adjustment of dials during operatoin. The Garrett *GTI 2500* uses automatic ground tracking or "Auto Track" in both Discrimination and All-Metal modes and also allows manual ground balancing while in the All-Metal mode to compensate for various ground mineralisation conditions.

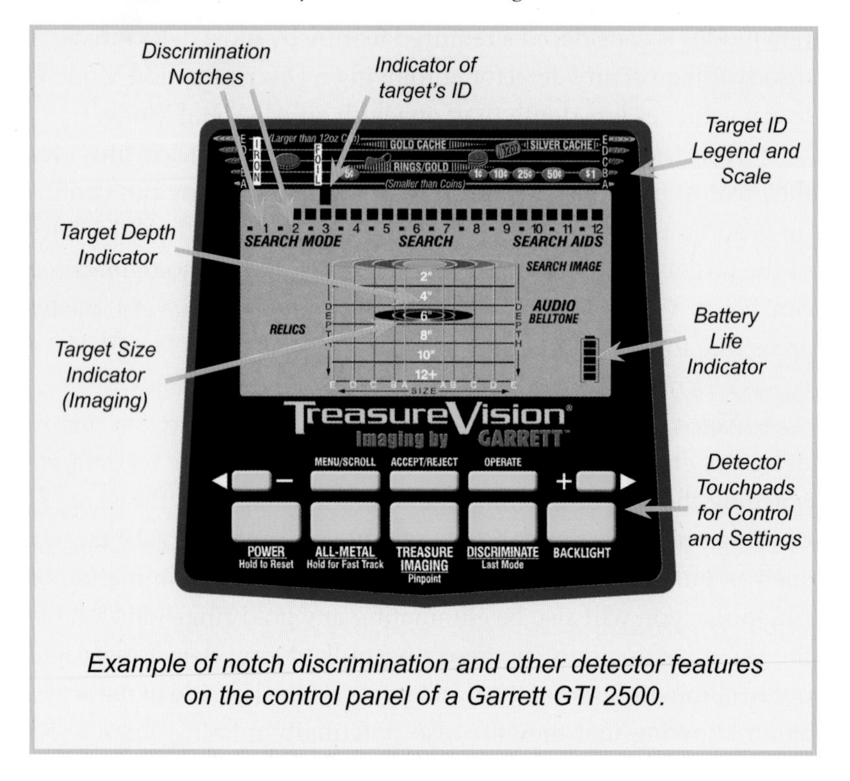

Example of notch discrimination and other detector features on the control panel of a Garrett GTI 2500.

• *Sensitivity*—This feature simply increases the depth of detection. Generally, you will want to get as much depth as possible as you search for older, deeply buried objects. Turning down the sensitivity can be beneficial in certain situations, however, such as the presence of power lines or other electrical interference; high ground mineralisation; or areas of high concentrations of rubbish.

• *Threshold*—Many of the less expensive detectors operate in what is known as "silent search" or a silent mode. As you step up into the higher-end machines, they frequently operate with a threshold tone which is used in the All-Metal mode. Although detectors with threshold adjustments can be turned down to silent while in All Metal mode, it is recommended that you always operate your detector with a minimum level of audible sound.

The use of threshold (also called "hum" or "tuning") allows you to hear the audio increase sharply whenever a desired target is encountered. With experience, you can also judge the size of a target and hear whether your detector is properly tuned or not.

• *Volume*—This allows you to set the volume to your preference according to the prevailing external environment (i.e quiet or noisy) conditions.

• *Variable Tone ID*—On some detector models, different types of metals produce different audible tones. Garrett detectors, for example, emit a distinctive Belltone ring whenever a target of high conductivity (such as a coin) has been located. Targets of lower conductivity, such as iron, create a sound with a lower pitch while targets of medium conductivity produce a higher-pitched sound.

• *Tone Adjustment*—This refers to the ability to adjust your detector's audio to a preferred pitch or tone. Increasing the tone will generally raise it into a higher treble range, while decreasing the tone will lower the pitch to more of a bass tone.

• *Imaging (target size and depth)*—This Garrett exclusive feature allows the user to view the size of a target and its indicated depth on the detector's LCD screen. Imaging indicates up to five different target sizes ranging from a item smaller than a coin to larger than a soft drink can.

• *Surface Elimination*—This feature is designed to let the detector ignore objects buried in the first two to five centimeters. In theory, you are only detecting the deeper and thus older metal objects while skipping over the modern trash items that are near the surface.

• *Salt Elimination*—When hunting on a beach or in any area with high salt content, this feature will help eliminate interference caused by wetted salt.

• *Coin Depth Indicator*—This meter helps you judge the depth of a coin-sized target. Large objects and extremely small objects will usually not be indicated accurately.

• *Frequency Adjustment*—Many higher-end detectors allow the user to operate at different detection frequencies to eliminate outside interference. This is particularly useful during rallies or in situations where you find yourself hunting in close proximity to other metal detectors. The change of frequency is minimal, designed only to prevent disturbing your search and does not affect target detection capabilities. A frequency adjustment option on a metal detector does *not* indicate that it is using multiple frequencies simultaneously.

• *Battery Level Indicator*—Some detectors will give an audible electronic signal indicating the level of the batteries. A detector with a graphic display or meter is more useful in the field; you can remain aware of the batteries' level and avoid being caught with a "dead" machine.

Purchasing a Metal Detector

Always try to *purchase a detector that best suits your particular hunting needs.* The places you will most often hunt and the targets you most often seek will help determine which metal detector is best for your needs.

• *Are you a gold prospector or a coin shooter?* The prospector will often hunt in areas of highly mineralised grounds where a Multi-Frequency (or PI) detector will be beneficial. The coin shooter will likely prefer a Single-Frequency Discrimination machine to help narrow potential desired targets from undesired targets.

• *Will you be working the inner city areas, freshwater streams or along the coastline?* The previous discussions on Single versus Multiple Frequency detectors cover some of the pros and cons of each detector type in relation to various environmental situations.

• *What are the soil conditions in the areas you will most often hunt?* People who hunt in mineralised soil might consider a pulse induction model and the use of a Double-D searchcoil to punch through the challenging terrain.

• *Are you planning on a specialized type of treasure hunting?* Gold prospectors and wreck divers should be aware of the demands required by the various styles of hunting. Books and articles can help. Because you will often be spending more than the person buying a general-purpose detector, your research will help prevent buyer's remorse.

• *Are you looking for a good, all-around detector to use during your travels?* If your answer is "yes," read reviews on detector brands that are in your price range and seek out a manufacturer with a good reputation for quality. The Internet is a great place to research and learn about detectors. The Internet can also be a confusing place to get biased opinions on detector brands because of the volume of discussion groups and sites that are driven by commercial interest. Be aware that some online advisors seek nothing more than to drive you to a particular dealer or importer. That said, there are plenty of quality detectors in each price range. One of the best pieces of advice therefore is to *decide which particular features are most important to you.*

• *Do you want a detector that is very lightweight?* Some detectors can become quite heavy after consecutive hours of searching. One helpful feature on some detectors is a detachable battery pack that can be mounted on your hip to alleviate some of this weight.

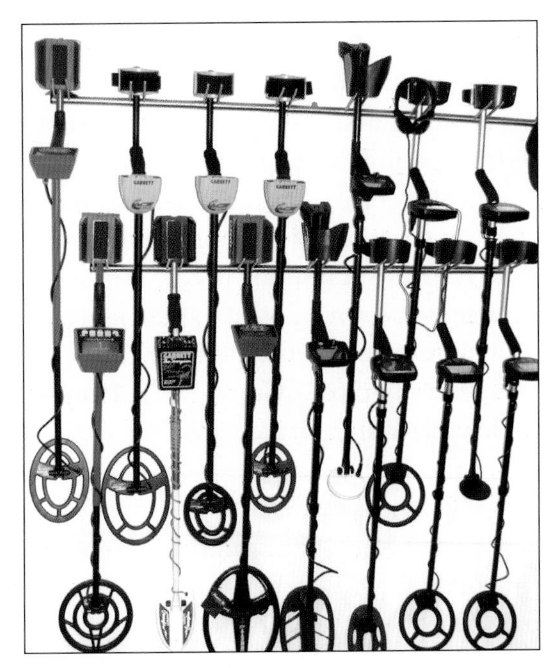

Your selection of a metal detector brand and model can be challenging with all of the choices today. Determine what features are important to you, check the manufacturer's reputation for quality and service, and evaluate your budget as you weigh the options.

• *How easy is changing the detector's batteries?* Batteries can run low in the field. Many detectors have battery covers that easily slide back for battery replacement, requiring no special tools.

• *Does the detector offer different searchcoils for hunting flexibility?* Most detectors are sold with a mid-sized searchcoil that is good for most hunting conditions. You might desire a Double-D coil for mineralised ground and also a smaller "sniper" coil for searches in tight spots. The more coils that are manufactured for your detector, the more hunting flexibility you offer yourself.

• *Is a training DVD included for quick learning?* The printed owner's manual included with your new detector is essential, but an instructional video provides an entirely different opportunity to understand your instrument. Many people can grasp a concept that is demonstrated on a video much faster than by comprehending the printed words in a manual.

• *What advice do other detectorists have for you?* Join a metal detecting club in your area and talk to people about their hunting experiences. But, always *consider the source.*

• *What is your budget?* For most people, there is a limit on how much money they can spend. Some new detectors can be purchased for less than 100 Euros from retail stores. There is also a saying that goes, "You get what you pay for." In other words, you will pay a little more for a detector of quality manufacture with excellent features. On the other end of the spectrum, there is no sense in spending thousands of Euros on the fanciest detector if it does not fit your hunting needs.

Decide *where* you want to hunt and *what* you hope to find. This will narrow down your detector choices. For the budget-minded, the modestly-priced Garrett *ACE 250* performs very well over a wide variety of environments.

CHAPTER 3

HUNTING TIPS AND TECHNIQUES

Philip Oyen took up metal detecting just three years ago and has become quite proficient. His Belgian detecting club buddies quickly decided that he had a lucky air about him. During a recent holiday, Philip and his friends traveled to Italy to search several areas on private property where they had the landowners' permission. One area was on a scenic hill overlooking the Mediterranean coastal town of Civitavecchia, located north of Rome.

Most of the group had some luck finding a few artefacts, tokens, a fibula and an occasional coin. Philip, however, drew friendly jeers from his buddies when they emptied their recovery bags at the end of the afternoon and he had produced 15 coins.

Philip started metal detecting with a Garrett *ACE 150* that he occasionally borrowed from a family member. In short order, he found two very nice silver coins from the mid-1500s in northern Belgium. He became interested enough in this hobby that he purchased his own *ACE 250* in 2008. With it, he soon found a gold French coin, dated 1515 AD from the rule of King Philip in the Middle Ages. Such gold coins are what many European detectorists only dream of finding. Philip is proof that while his years of hunting are fewer than that of some of his friends, he was quick to learn his own detector and to pick up good techniques.

His friends call him lucky but Philip just laughs and in his own good-natured way, he chides back with, "Luck? Ahh, you're just jealous of me!"

Philip Oyen *(left)* hunts a coastal Italian field with his Garrett *ACE 250*. Although he has only been metal detecting for three years, he has become quite skilled.

(Below, left) Philip found these two 1500s-era silver coins with an *ACE 150*. After buying an *ACE 250*, he found this 2,500-year-old Roman silver ring *(below, center)* and this French gold coin of King Philip, dated 1515 AD *(below, right)*.

How to Use Your Metal Detector

There is nothing wrong with good luck, but you must first understand your detector. If you have just acquired a new detector, start by reading its manual and watching any DVD that was included. The more you learn about your metal detector the more effective you will be at locating coins, jewelry and other treasures.

Start at home with “air tests” by passing various treasure targets in front of your coil to familiarize yourself with where exactly they “hit” on your detector’s target ID scale. It is great practice to create you own “test plot” by actually burying several targets to see how they sound in the soil and where they register on your target ID scale. This should include several coins or artefacts as well as a bottle cap, a pull tab, and iron junk metal. Bury each item about four inches deep, making sure to space them a good foot or so apart. Create a map of each item’s location or use some marking system to keep track of the precise location of each target item. Practice scanning each of the targets while listening to and studying all of your detector’s detection signals.

Make sure that you adjust your detector so that the searchcoil can swing comfortably just above the ground's surface. If your stem length is too short, you will find yourself leaning forward slightly and you will end up with a sore back. Stand up straight and extend your arm naturally while holding your detector. You will want to move the searchcoil slightly forward of your body and adjust the stem until your searchcoil hovers just above the ground. Hold your detector with your dominant hand, the one you write with. You can switch back and forth if one arm gets tired during a full day of hunting, but most detectorists prefer to swing their machine with their dominant hand.

This leaves your less dominant hand free to carry your shovel or digging trowel. When you prepare to dig a treasure signal, you can use the less dominant hand to dig in soft soils while continuing

CORRECTLY ADJUSTED STEM LENGTH
This detectorist is able to maintain good posture with his searchcoil extended before him and hovering just above the ground. He will remain more comfortable after long hours of searching.

STEM LENGTH TOO SHORT
This detectorist will soon have a very sore back from having to lean forward in order to keep his searchcoil hovering just above the ground.

to hold onto your detector with your dominant hand. Where you are able to do this in fields with softer soil conditions, your target recovery time will be improved. You will find it more tedious to lay your detector down, dig, pick up the detector to rescan the excavated soil, and so on. Of course, some soils are so rugged that you will be unable to dig without setting your detector down.

Tips on Proper Scanning

Turn on your detector's power while holding the searchcoil a foot or two above the ground. Do not turn on the power while your coil is in close proximity to metal. Your detector may attempt to "tune it out" or discriminate this particular metal type. After powering your machine on, lower the coil and you're ready to begin hunting.

Keep the searchcoil level as you scan and always scan slowly and methodically. Scan the coil from side to side and in a straight line in front of you with the searchcoil about 2 cm above the ground. Do not scan the searchcoil in an arc unless your arc width is narrow (about 60 cm) or unless you are scanning extremely slowly. This preferred straight-line scan method allows you to cover more ground width in each sweep and permits you to keep the searchcoil level, especially at the end of each sweep. This method reduces skipping and helps you uniformly overlap the areas you have scanned.

Level swings help reduce the degrading effect of ground mineralisation. Because of its conductive and magnetic minerals, the ground always creates a response that can reduce your detector's accuracy and detection depth. It is therefore very important to maintain a level swing with a consistent searchcoil height to achieve optimum performance. Where the ground varies in elevation or has a rugged contour, your searchcoil should follow the rises and falls of the ground at a consistent height.

Overlap by advancing the searchcoil approximately 50% of the coil's diameter at the end of each sweep path. You want to sweep an area

Veteran detectorists are able to recover targets from many fields without laying their metal detector down. *(Upper left)* This Belgian detectorist uses his left hand to dig a big scoop of soil from his pinpointed area. *(Upper right)* He then scans the excavated soil to determine if the target has been removed or is still in the ground. If the signal remains in the hole, he will take another scoop out with his shovel and repeat the process as necessary.

thoroughly because depth penetration is less at the very edge of your searchcoil than in its center. Occasionally scan an area from a different direction, particularly if you get an erratic or suspicious target signal. Unless you overlap your swing, you will skip patches of ground as you sweep forward. For smaller sites you intend to work over thoroughly, cross back over the area from a different direction and see what your detector finds this time.

Be careful not to raise the searchcoil above normal scanning level at the end of each sweep. When the coil begins to reach the extremes of each sweep, you will find yourself rotating your upper body to stretch out for an even wider sweep. This gives the double benefit of scanning a wider sweep and gaining additional exercise.

As you scan the searchcoil over the ground, *move the coil at a rate of about 30 centimeters per second.* Don't get in a hurry, and

POOR SEARCHCOIL SWINGING TECHNIQUE will result in decreased treasure discoveries. This animated photo series illustrates why you should always swing your coil level to the ground. You will pick up the deep coin shown in the middle but targets at the same depths on the outer edges of your swing area will be lost if you allow the search head to swing upwards.

CORRECT SEARCHCOIL SWINGING TECHNIQUE. In this animated photo series, the same metal detectorist is able to pick up all four coins at the same depth by using good form in his searchcoil swing. Always keep the search head level with the ground surface to achieve the maximum depth while searching for targets.

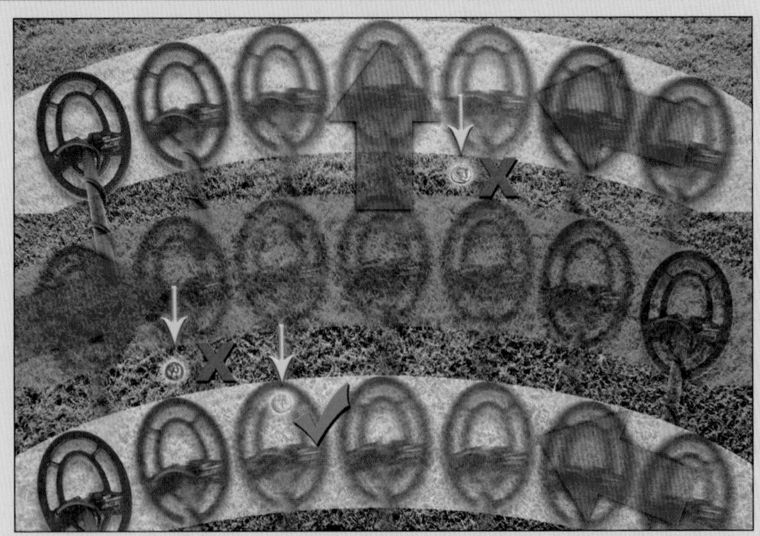

CORRECT OVERLAPPING OF YOUR SEARCH COIL SWING will prevent skipping of good treasure targets. The illustration *above* depicts a detectorist who is advancing while swinging the searchcoil back and forth, creating patches of ground that are missed by the coil. Two good coin targets have been missed by not overlapping the searchcoil swing.

(Below) This illustration shows the same detectorist overlapping his swing by about 40% of the searchcoil's diameter as he moves forward. This method will slow your advance but prevents you from missing good targets in the gaps created by the method shown above. Use a larger searchcoil to cover more ground as you advance. All three coin targets would be detectable in this method.

don't try to cover an acre in ten minutes. In your mind, you should always imagine that what you are looking for is buried just below the sweep you are now making with your searchcoil.

Target ID Tips for Discrimination Detectors

In Chapter 2, the science of target identification with Single Frequency Discrimination detectors was discussed. The target metal's conductivity, permeability, thickness, size, shape and orientation in the ground all play a role in determining where the item will register on your Target ID scale.

More than two dozen countries in Europe have traded in their traditional currency for the Euro. This new coin can be found in abundance across the continent on beaches, in parks, in the field

Your detector will report the various Euro coins differently due to their composition. The smaller 1-, 2- and 5-Euro cent coins are made of coppered steel, which is easily lost when your detector is set to discriminate iron. The 10-, 20- and 50-Euro cent coins are made of Nordic gold, an alloy that allows them to register solidly on your detector. The larger 1- and 2-Euro bi-colour coins will register consistently on your detector.

1
DEUTSCHE
MARK
10
PFENNIG
BELGIQUE
10 F
5F
BELGIQUE
20
25
DANMARK
1
20
1
SCHILLING
REPUBBLICA ITALIANA
L.500
JULIANA KONINGIN DER NEDERLANDEN
10
CENT

and within cities of every size. It is worth noting that not all Euros are made of the same metal, making them register differently on your metal detector's discrimination scale. As of mid-2009, there were more than 84 billion Euro coins in circulation worth close to €20.7 billion.

The smaller 1, 2 and 5 cent Euro coins are the most erratic for your detector's target ID scale. They are made of a reddish-colour coppered steel. Metal detectorists who discriminate out iron might only faintly detect these low denomination coins. The 2-Euro coin is slightly more conductive but can be missed if you opt to discriminate out pull tabs.

In dry sand and regular soil conditions, you can employ more discrimination and still easily detect Euros of 10-cent and higher values. The 10-, 20- and 50-cent Euro coins are made of what is called *Nordic gold*, an alloy that was originally developed for the Swedish Mint. Nordic gold is a brass alloy composed of more than 89% copper, 5% aluminum and 1% zinc. These coins will sound solidly on your detector and register mid range on your scale.

The more easily detected 1- and 2-Euro coins are made of an alloy of nickel, brass and copper. They are known as *bi-colour coins* because of their silver-coloured outer ring and gold-coloured inner part. While the 1- and 2-Euro coins will register as a very solid hit when they are lying flat in the ground, they will jump around a little on your detector's scale when they have an orientation in the ground standing on end.

Since the introduction of the Euro, the currency previously used by various European countries is becoming collectible for some detectorists. In many of these countries, it is already too late to exchange such "old" money for modern Euros. Austria, Germany, Ireland and Spain have not yet restricted their final dates to exchange disused currencies. By the end of first quarter 2012, Finland, France, Greece and Italy will no longer exchange old money for Euros.

Seen in this detectorist's collection are coins from various European countries which have switched to the Euro. From top to bottom, these coins are from: Germany (top row); Spain (second row); France (third row); Belgium (fourth row); Denmark, Czech Republic, Poland and Austria (fifth row); Italy (sixth row); the Netherlands (seventh row); and Austria (bottom row).

Target Masking and Discrimination Tips

Iron is a problem for many detectorists and their machines if they are hunting particularly trashy areas. The grounds where ancient Roman dwellings once stood can be littered with small pieces of surface trash metal such as iron nails, bolts and shards of metallic rubbish of all shapes and sizes.

In such areas, you will often choose to use some form of discrimination with your detector. The problem is that you will also be eliminating some potentially good targets by discriminating iron. Old rubbish dumps, homesteads and gardens are also often littered with small pieces of iron. Such objects create a camouflage effect on the good treasure targets that lie below. Your detector often cannot see through the junk metal that is masking a good target underneath. This camouflaging effect is known as target masking. Some detectorists also use the phrase "iron masking," but since this term in used on some detector brands to describe their iron elimination mode, the term "target masking" will be used in this text.

Experienced detectorists investigate target signals carefully from multiple directions. The sound of one good target signal that is accompanied by the sound of junk metal is often worth another few seconds of time. Sweep the searchcoil back over the area from other directions, walking around the target area in a circle. If a good target is partially masked by iron or is lying vertically in the ground, the target response might be very limited. A hazard of ignoring weak or irregular signals is that good targets that are deeply-buried can become inconsistent near the limits of detection depth.

Target masking is certainly an issue in the field, but there are solutions to help overcome these situations. **Among the target masking solutions you can test:**

- *Use a smaller "sniper" searchcoil for areas with heavy concentration of iron or other trash metal.* The smaller size coil will examine smaller spaces where good coins might be waiting between pieces of junk metal. Smaller sniper coils also allow your detector to

SWEEP TECHNIQUE TO OVERCOME TARGET MASKING

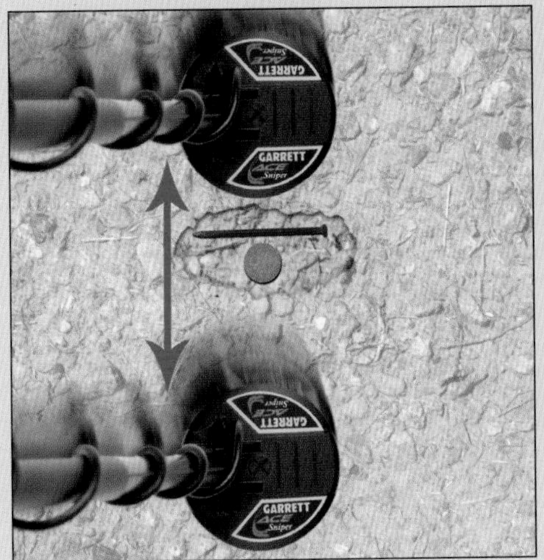

FIGURE 1

ALTER THE DIRECTION OF SEARCHCOIL SWING to help overcome the target masking effect. As depicted in these illustrations, a good coin target laying adjacent to iron might be largely masked from one coil sweep direction.

(Figure 1) The detector may give alternating signals (unable to separate the two targets) or only an iron signal (unable to determine that a good target is present) when the searchcoil passes over the coin and nail from this direction.

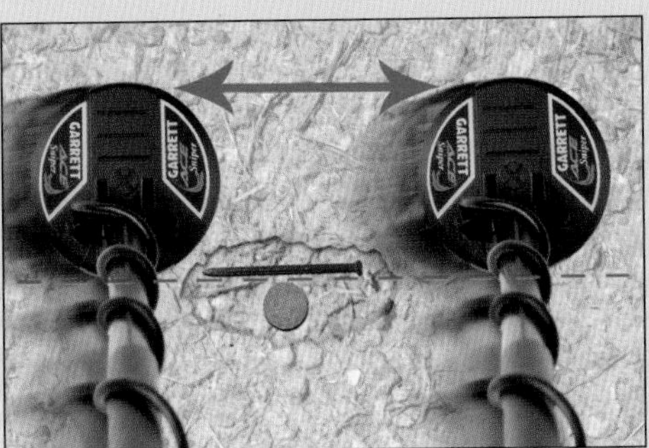

FIGURE 2

Walk around the unknown target (about 90°) and then swing the searchcoil from a new direction to check target response. As the searchcoil moves forward it passes over the nail only *(see Figure 2)*, and the Target ID gives an iron reading.

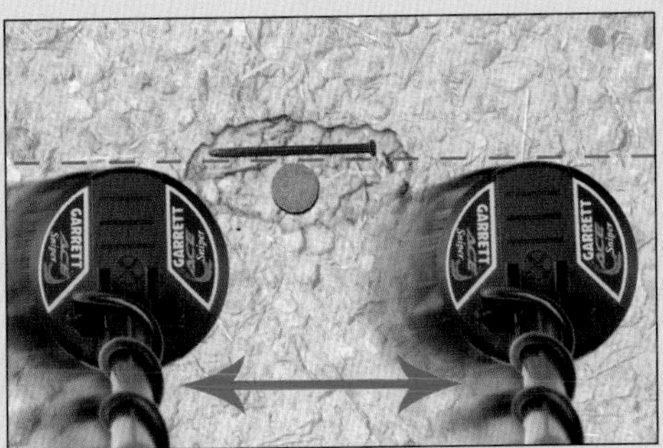

FIGURE 3

Continue sweeping over the target area and move the coil back toward the lower edge of the target response area. As the coil moves away from the iron nail and over the coin *(see Figure 3)*, the metal detector now offers a good signal from the coin. By physically separating the conflicting signals, the detectorist is able to determine that a good target is present.

search close to metal fencing, sidewalks and foundations where rebar can negatively affect larger coils.

- *Turn down the sensitivity of your detector.* Although your fear in doing so is losing detection depth, you will also be decreasing your detector's reaction to the iron in troublesome spots.
- *Use your test plot to learn more about the effect of target masking.* Bury several good targets at about six inches deep. Then, plant nails and rusty iron targets above your target. Practice sweeping over the good target in All Metal mode and then with various discrimination settings. You will find the good target harder to find in a discrimination mode, and it might not register at all.

You can also conduct air tests to see how target masking affects your detector. Hold both a piece of iron and a silver treasure target in the same hand (making sure there are no other metallic objects on that hand). Pass these two items together in front of your searchcoil in a discrimination mode and then with your detector set to All-Metal mode. Can your detector pick up the good silver target at all?

The combined effect of these two metals will produce varying results on varying detectors. Some will see the iron item and, in an Iron Discrimination mode, will not signal a target at all. This is generally the case with a detector that collectively notches iron items under one segment. When this notch is rejected, the detector then rejects any and all iron targets.

Other detectors with more selective iron discrimination will produce some type of target signal. This might be an erratic cursor that bounces between notches because of the two different types of metal being in close proximity. Higher frequency detectors generally are better suited to overcome such target masking but advanced/adjustable iron discrimination generally must also be present.

The iron object and the silver target each possess different levels of conductivity that register as completely different target ID patterns when they are scanned separately. When these two items are placed together, a new combined conductivity has been cre-

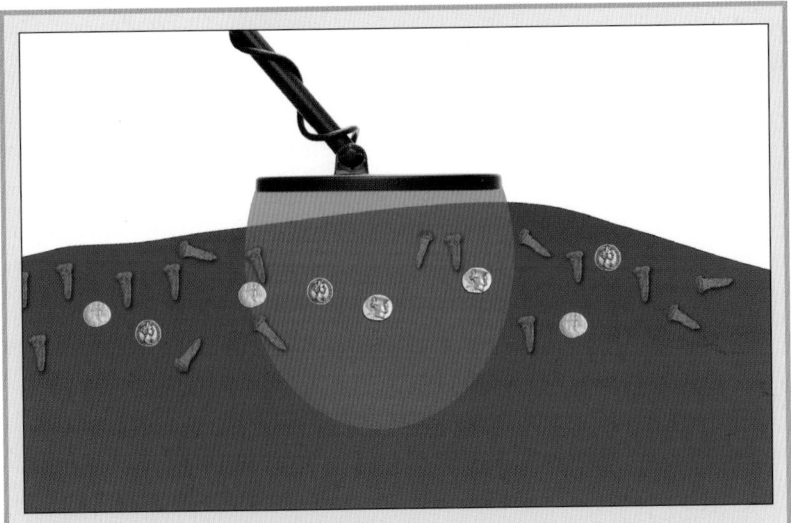

TARGET MASKING can be caused by several issues. In the above scenario, the detectorist is utilizing a larger searchcoil and employing iron discrimination. This old townsite contains a number of iron square nails with valuable gold coins mixed in with the iron debris. The effect of "iron masking" is preventing this large coil from seeing the good targets.

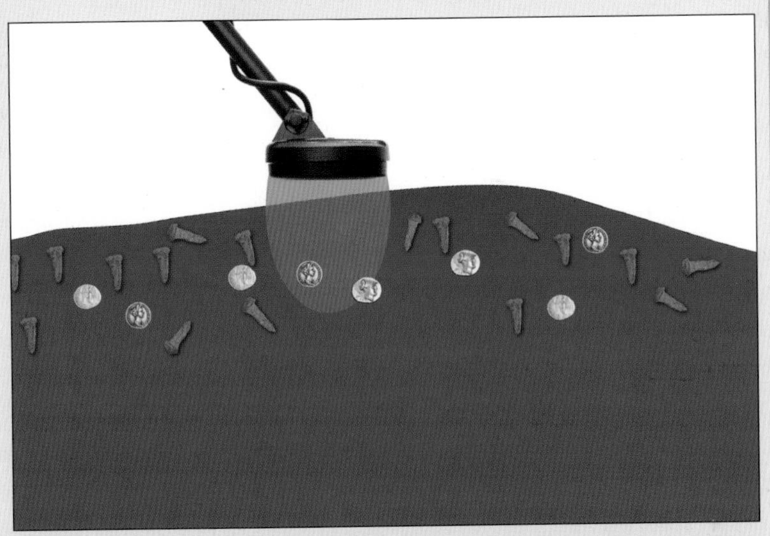

THE USE OF A SMALLER "SNIPER" SEARCHCOIL can help in such trashy areas to allow good targets to be seen when they reside in close proximity to iron. Such smaller coils also allow you to search closer to metal fencing, foundations and sidewalks.

ated. The result is that the pair now register on a target ID scale, or produce an audio response, that corresponds to something in slightly higher conductivity than iron.

Be aware that on some metal detector brands, *any target more conductive than iron will produce a "good" target sound*. This does not guarantee a good target will be unearthed. Iron and aluminum, for example, will combine to produce a reading that is a little better than iron. The advantage of a detector with notch discrimination, such as an *ACE 250* or *GTI 2500*, is that target readings are more precisely identified. The disadvantage of notch discrimination is that these segments of discrimination offer less control. If a notch is rejected, all items within this range will be ignored. Metal detectors with older style technology (i.e. dial controls) do allow tuning that permits determining where to drop out iron items.

Try placing the iron and silver items on the ground in very close proximity. Use your All-Metal mode and then discrimination modes to sweep around these two objects. Take note of how the iron item affects your identification of the silver item as you approach the target from different angles with your searchcoil. Try switching from a large searchcoil to a smaller, sniper-sized coil to determine how this can improve your ability to detect the good silver target.

- *Utilize a slower search speed with shorter swings in an iron-heavy area.* Your metal detector's "recovery" speed is slightly longer after it encounters a negative item such as iron. Anytime a metal detector encounters a target signal, there is a certain degree of time before the machine recovers. Detectors with quicker recover speeds present lesser risks associated with iron objects masking out good treasure targets. If you are scanning an area rapidly where your detector signals iron items, you may be passing over good items in the split second before your detector recovers. Some detector models offer variable scanning speeds, but in general *slow down* when you are working against iron rubbish in a good spot. Shorter-width swings also reduce the number of targets your searchcoil will encounter.

• *Learn to recognize the false signals that non-motion detectors will occasionally emit.* Ground mineralisation is one of the key causes of this effect. You can determine a false target by making repeated swings with your coil over an area that is giving erratic or short sounds. A good target will produce a solid, repeatable response on most instruments. Non-motion machines will detect a metal target even if the searchcoil is hovering over the target without being moved.

Sometimes, a true target that is very small or deeply buried may only produce a good signal from only one direction. This is why veteran detectorists often explore suspicious signals by circling the spot and swinging their coil from different directions to seek a repeatable, solid signal. Again, be aware that even good targets can become inconsistent if they are deeply buried near the limits of the detector's depth capability.

• *Test your detector's recovery speed by conducting an air test with a coin.* Pass the target back and forth in front of your searchcoil to see how well it continues to offer a clear target signal at varying speeds. Muddled audio responses let you gauge your detector's recovery speed.

• *Use a depth multiplier to find larger, deep targets without the interference of surface "junk" metal.* Such two-box configurations are operated in All-Metal mode and are ideal for deeply-buried items of moderate to large size. Small, shallow objects which can camouflage good items with a standard searchcoil are simply not detected. The Garrett *TreasureHound Depth Multiplier* with *Eagle-Eye*™ Pinpoint, however, includes a smaller front coil for accurate pinpointing of smaller objects with the touch of a button.

• *In an important area with high junk concentration, use the All Metal mode of your detector to first find and dig out the iron trash.* This is, of course, more time-intensive and is probably your most unfavorable choice. An important archeological site that must be thoroughly searched might require such measures. Once the surface trash in cleared, however, the camouflaging effect of the iron will open up the ground for the detector to seek any good items below.

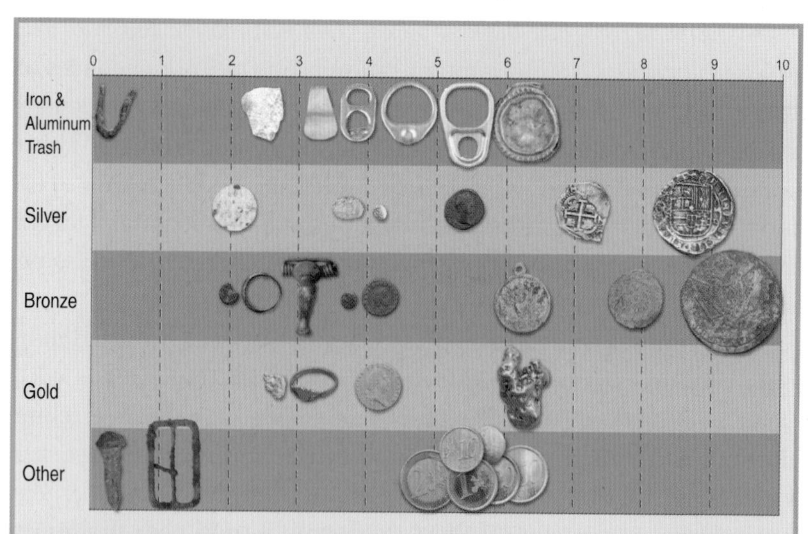

Sample gold items (seen above).

EUROPEAN TARGET CONDUCTIVITY of sample items detected in All Metal Mode. This chart illustrates the fact that conductivity is only one factor in determining where a target will register on your detector's ID scale. For example, the silver targets shown above vary based on their size and thickness. Be aware that these same targets can also register differently based upon their orientation in the ground.

Junk metal items (seen above).

Sample silver items (most shown in chart above). Note how size and thickness affect Target ID.

Sample bronze items (some seen above).

Euro coins and musket ball (seen above).

DISCRIMINATION OR ELIMINATION?

Use the target ID scale on the facing page for reference as you study the illustrations below. Once discrimination settings are applied in the examples below, the detectorist begins to lose good targets along with the junk items.

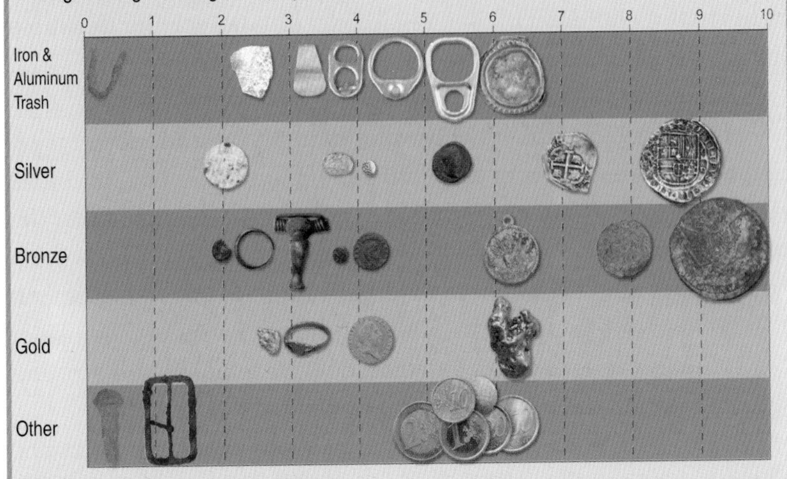

Example 1: IRON DISCRIMINATION
In the illustration *above*, the detectorist has elected to eliminate iron. (Depending upon the detector used, this might be accomplished by using Accept/Reject notches or by using a "Relics" setting.) The ancient square nail artefact has also been eliminated in the process (along with the rusty fence post nail). Note the shaded items which have been eliminated.

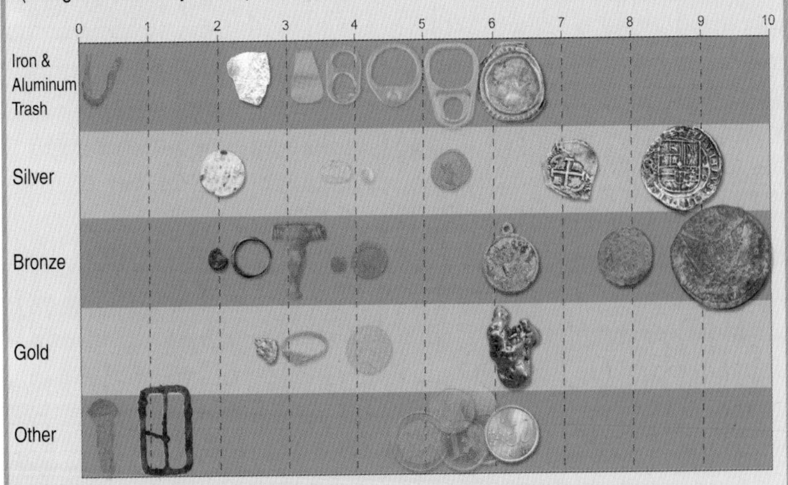

Example 2: HEAVY DISCRIMINATION
In this illustration, iron, pull tabs and ring pulls have been eliminated. By discriminating these items, gold rings, certain artefacts and some coins have also been eliminated.

General Metal Detecting Tips

The following techniques are basic recommendations that can benefit all metal detectorists, regardless of whether the treasures sought are coins, jewelry, caches, artefacts or something else.

- *Use headphones with your detector.*

Veteran detectorists have learned that the faintest, deep target signals are often hard to hear with the naked ear. This becomes especially true if you are hunting in a windy area such as on the beach. You are better able to concentrate on your search with headphones since you ignore external sounds. Your machine will also be silenced to people around you in public areas. The use of headphones can extend your detector's battery life by not sending signals through the built-in external speaker.

There are a wide variety of commercially-available headsets. Consider the areas in which you will hunt as you contemplate the headphone features. Select headphones with adjustable volume control and a long, spiraling flex cord. This stretchy cord helps prevent your headphones from being pulled from your head as you are digging a target. Those headsets with thin, straight cords can easily become snagged on tree limbs, vines and other outdoor obstacles which can damage or even snap the wires.

Wireless headsets can be used to help prevent such tangling of wires and the nuisance of your head being forever attached to your metal detector as you dig targets.

- *Use discrimination to eliminate only the targets that you are willing to bypass.* With a true All-Metal, non-motion detector, you will obviously detect all types of metal. The advantage of a discrimination detector with Target ID capability is that you are able to focus on the metals you wish to detect and ignore the metals you do not want to detect. The risk that comes with this metal discrimination advantage is that certain good targets are potentially missed.

For example, an iron-discrimination mode will prevent many rusty metallic items from registering on your detector. Certain interesting artefacts, such as Roman square nails or tools, will be discriminated out at the same time. Pop tops, pull tabs, bottle caps and aluminum foil are all bitter enemies of the metal detectorist. Here again, there is a danger inherent with simply notching out these items with accept/reject controls. Certain gold rings, fibula and coins register just the same on target ID scales.

It is up to you to determine how much discrimination to employ based on what you hope to find. Many European detectorists are only willing to eliminate the first one or two notches on their discrimination scale for fear of missing good, low-conductive artefacts. *(See the illustrations on the preceding pages.)*

- *Field test a new detector at a location that has been productive for you in the past.* Those with improved technology or even using different searchcoils might afford you deeper detection depth or better discrimination than detectors you have used in the past.

Gilles Cavaillé, for example, has a personal testing ground for new metal detectors that he has used for many years. When he

Gilles Cavaillé has recovered 500 gold coins from a piece of property in southern France during a 37-year-period. Ten of these coins are shown above. (Left) This is the most recent gold coin found by Gilles in 2007. He opted to keep it on display still in the clod of earth which he dug from the ground.

was ten years old in the early 1970s, his parents purchased a new property in southern France. During the final stages of home construction, his parents hired a man with a bulldozer to level the soil so that they could lay down new grass. Gilles was watering the new grass one day when he happened to spot a shiny, gold object. It was an 1812 gold Napoleon coin of great value.

Years later, when Gilles had become interested in metal detecting, he began searching his parents' property for more of the gold coins. To his great surprise, their 30 x 200-meter tract of land has proven to be productive. In recent years, he has taken various brands of detectors to search this property to see how well they can find more of the coins. On one occasion while helping his father plant a new tree, Gilles found another of the gold Napoleon coins. Using his metal detector, he scanned the deep hole for the tree and found a total of 45 gold coins in one afternoon!

The land was once used by sheep farmers and Gilles can only imagine that some early settler hid a large cache of these gold coins, which have become scattered over nearly two centuries. "I found my 500th gold coin on my parents' property in 2007," he says, "but I'm still hopeful that I will find more in the future. Any time I have the chance to test out a new metal detector, this is my special testing ground. As technology continues to improve, I hope that I will find more gold coins here."

- *Prepare for the ground conditions where you will search.*

Most manufacturers are shy about offering an answer to the question of, "How deep will this machine detect?" There are really just too many variables. The size of the searchcoil, ground mineralisation, moisture, ground compactness, target metal conductivity, detector design and frequency, sensitivity settings and skill of the detectorist are just some of the many factors that come into play concerning how deep your detector can penetrate in a specific hunting environment.

If you are planning on hunting the sea coast, you should be aware that wetted salt sand generally causes problems with Single-Frequency (VLF) detectors. Many detectorists opt for a

PI detector for hunting the surf and the wet beach sand. Single-Frequency motion detectors with salt elimination modes, however, can be effective.

Another challenging area is any geological area of past volcanic activity where there is a heavy distribution of iron-laden minerals such as magnetite, laterite and hematite. *Black sand*—an accumulation of slightly magnetic heavy minerals most often encountered in mountain-area streams—is primarily composed of pulverized magnetite. Black sand can cause interference with VLF detectors and a loss of depth. In rare instances such as on some beaches in Italy, detectorists can be challenged with both salt water and black sand combined.

Again, Multi-Frequency (or PI) detectors excel in these adverse conditions. On Single-Frequency motion machines, detectorists should adjust their sensitivity settings to help cope with such negative effects of mineralisation or lower their discrimination levels. Another Single-Frequency remedy is switching to a smaller size searchcoil. Switching into the All-Metal Mode will generally restore lost depth, but it makes for a more difficult hunting style, especially for the less experienced user.

Inland hunters are equally challenged at times with the effects of mineralised soil. High concentrations of iron oxide or granular particles of the previously mentioned black sand minerals are encountered at times. Such conditions can lead to false signals and greatly reduced detection depth. Mineralised stones containing magnetite or other conductive elements are known as "hot rocks," and they can also create havoc with your detector. To overcome the effects of hot rocks, many detectorists utilize Multiple-Frequency detectors or lower their discrimination and/or sensitivity.

In the absence of heavy mineralisation, *damp ground improves conductivity and search depth.* Therefore, a productive ploughed field may yield even deeper targets after a few days of soaking rain showers. Conversely, freshly ploughed fields of inconsistent dry soil reduce detection depth. Your discovery yields should improve after allowing ploughed ground to settle for at least several weeks.

HUNTING TECHNIQUES: Retuning Your *Pro-Pointer*

The Garrett *Pro-Pointer* can also be retuned to pinpoint smaller, elusive targets. By "re-tuning" your pinpointer to your target, you are changing the audio field to report only the peak audio response of the target.

1. Slowly scan toward the target until the *Pro-Pointer*'s response increases to the full/constant alarm.

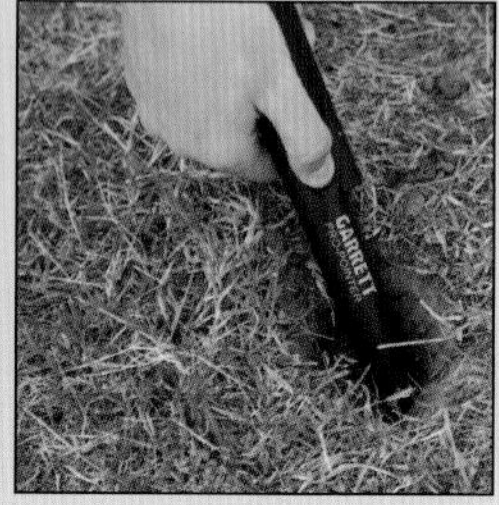

2. Then, without moving the detector from the target, switch the power off and then quickly back on again in order to highlight only the peak audio response of the target.

3. Now, continue scanning towards the target to find its precise location. Repeat this power off/on cycle to further narrow the audio response as needed. After you have finished, you can return the *Pro-Pointer* to its normal detection by simply switching the power off and back on again while holding it away from all metal.

- *Retune your searchcoil while pinpointing to tighten your search.*

One lesser-known trick can be used on some detectors to help you pinpoint smaller, elusive targets. Metal detectors which have a Pinpoint or Treasure Imaging button are using an All-Metal mode when you depress this pinpoint function. By "retuning" your searchcoil to your target, you can change the audio field to report only the peak audio response of the target.

Such retuning does not change the size of the detection field that your searchcoil is generating. Instead, the rise and fall of the audio response field has been narrowed to highlight only the peak audio response. In short, you must move your searchcoil much closer to the direct center point above the target object in order to generate an audible response.

This retuning process is quite simple for detectors that have a pinpoint button. First, pinpoint your target as you normally would by holding down the pinpoint button and maneuvering your coil above the target to the point where you receive the strongest target response signal. Then—while continuing to hover directly above this target—quickly release the pinpoint button and immediately press the button again and continue to hold down the pinpoint button. By doing so, you will have retuned the audio field against the target metal to generate only the peak audio response. As you continue to hold down the pinpoint button and carefully position the coil above the target area, you will find that your signal is coming from a much tighter pinpoint area.

Again, you have not altered the detection field but merely the audio field. Visually, however, it will appear as you move the searchcoil over the target area after retuning that you have cut the detection area in half during this special pinpointing process.

This retuning method obviously requires some practice. Place a coin on the ground and practice until you can clearly see how the concentration of the target's peak audio response allows you to pinpoint the item more precisely.

• *Search old homes, including attics and basements.*

Metal detecting can be quite productive in older homes that have passed from owner to owner with time. Bruno Lallin of Paris enjoys searching through attic spaces of early homes, the dirt floor cellars of castles and in ancient barns and garages. He has made great recoveries of cached money and valuables that long-deceased former owners secreted away.

Sometimes old wooden floorboards can hide a box of hidden money or other valuables. Bruno has been called upon by relatives

(Above) Barns, attic spaces and basements, such as this old French homestead, can yield great sums of hidden loot and other valuables. Courtesy of Bruno Lallin, Lutéce Détection.

to help search for gold coins and money caches that the relatives were unable to locate after their loved one passed away. (For more on cache hunting tips, see Chapter 6.)

Finding Productive Detecting Areas

The most obvious question that comes to mind for the new detectorist is, "Where should I start searching for good finds?" Truth be told, old coins and nice artefacts can be found all across the European continent. Veteran detectorists rely upon research, information gleaned from fellow detectorists and their own scouting efforts to locate areas that might prove to be productive.

- *Conduct an area "recon" of your site.*

Before you begin detecting, you should do a reconnaissance scout with your eyes to look for clues. Broken bits of pottery in a field are great indicators of an old homestead of village. If you have studied enough pottery shards to know much about them, they can also be key indicators at to relative age of items you

Nigel Ingram snorkels through the ruins of a Roman village in Turkey that collapsed into the ocean many years ago following a powerful earthquake. Your ability to spot potential hot spots improves with your familiarity of Roman, Celtic and medieval building materials.

expect to find in this field. This is not to say that you should walk the whole field looking for clues before you start detecting; keep a keen eye out for visual items at all times. Many detectorists have spotted brilliant artefact items, churned up by ploughs, before they even came within range of their searchcoil.

Study the detail of ancient Roman or medieval structures. There are places where the ruins of such structures are viewed by thousands of tourists each year. Rome is a prime example where ancient architecture can be studied. The pottery fragments of ancient civilizations in Europe tend to be greyish brown or gritty black in appearance. This is in contrast to more recent pottery shards from the medieval times to more modern centuries in which the pottery can be glazed or more brightly coloured or elaborately patterned.

Take note of the height of plants and vegetation in areas where old settlements were once located. The colour of the soil will look different where it has been mixed with layers of old rubbish. Look for long rows of old trees that don't seem to have grown in such a

A group of detectorists received permission from the farmer to scout about his cornfield in France for signs during the summer. They found bits of broken pottery *(left, inset)* in the field, which gave them hopes that this could be a worthwhile field to search after the crops had been harvested later in the year.

formation by pure chance. These might lead you to the remnants of an old foundation, wall or collection of large stones that were once part of a homestead, castle or early structure.

Oysters, snails, clams and marine mussels were a part of the ancient Roman and medieval diets. Many detectorists have learned to look for an abundance of discarded shells as an indication of previous habitation. Unusually large groups of overturned rocks in a field where rock is only lightly scattered might indicate the remains of an old structure. You should also watch for small areas

The loose soil of a freshly ploughed field needs time to settle into a more consistent state in order to achieve greater detection depths.

in a ploughed field where the soil is notably lighter or darker in colour than the rest of the field. These can all be signs of a deserted medieval village or a Roman settlement.

Finding such signs does not guarantee that your metal detecting efforts will be successful but they certainly don't hurt your odds of finding a hot spot.

Once you have located such promising signs, conduct a search of the area to pinpoint hot spots. For a large field, you can sweep the entire perimeter of the area in hopes of locating a crossing point. Then, work two diagonal sweeps across this rectangle in the form of an "X" shape through its center. This is obviously not to be considered a detailed search of a large area but it is a quick way to sweep a zone without spending too much time. If you find a nice coin or artefact, then you can conduct a more thorough grid search by walking lanes up and down through the area.

- *Scout for other search sites in your spare time.* Look for any evidence of habitation by walking the fields during the "off season" when crops do not allow you to hunt a farm. Europeans often refer

Henry Tellez *(left)*, Steve Moore and Nigel Ingram had a good afternoon's hunt along this old right-of-way path that cuts through various fields in an English countryside. Such public paths (note the greener grass color along the path) have long been used for pedestrians.

(Below) Steve holds up a George III half-penny found near this public path. The trio, assisted by Nigel's dogs, unearthed several British coins, a Roman coin, buttons and other artefacts in their search.

to this as *field-walking*. In rural areas of England, privately owned land is still required to contain public right of ways through them. Local settlers may have walked across the fields to a school or church for many hundreds of years. The landowners in these areas put up a special gate if they have livestock that might escape. You are not allowed to metal detect along these public paths without obtaining permission from the land owners. Field-walking along these public right of ways, however, can help you spot clues which point you to a potentially good hunting area. You can then contact the owners of these fields that appear to be interesting.

- *Exercise courtesy when approaching the landowner for permission to detect.* Start by explaining why you think their land might contain some historic items. Show or describe your research efforts to reveal your knowledge of the area. The property owner might be interested to learn history of their property which they were not aware of.

Think about the best time to approach a landowner to seek permission. For example, if you know upon a farmer's door in the

HUNTING TECHNIQUES: Tips from a European Detectorist

Graham S. of Scotland (left) has hunted for three years with his *ACE 250* metal detector. "I find it to be a great value machine," he says. "I detect mainly on pasture and stubble. When the ground is wet, I find it to be an even greater machine.

"In some cases, the ground in Scotland can be very mineralised. There are some things you can do which I have learned from the forums. For example, if the signal is 'iffy,' I will pierce the ground with the trowel and then sweep over it again to hear how it sounds and also adjust my sensitivity."

(Above) One of Graham's two favorite recoveries is this Scottish King Charles I forty pence silver coin, which was produced between 1637 and 1642.

(Left and right) Graham's other favorite coin recovery is this King William III silver shilling from 1696.

evening you might find him to be less than open-minded to your ideas. You should realize that the farmer could have been working his fields, livestock or property since long before dawn and is likely preparing for bed. You would be better served to meet with the farmer during the morning hours when he is not exhausted from the day's chores.

Be courteous to the landowner regardless of the answer you receive. A "no" can sometimes be turned into a "yes" if you are polite. I approached some landowner in the U.S. about hunting a battlefield that I suspected to be on their farm. Before asking about metal detecting, I first explained my interest in history, my previous writing experience and my desire to prove some historically important points about the battle. They knew of the battle, and we talked for a while about what they knew of it and what kinds of artefacts their grandparents had found in the past. Eventually, I brought up the fact that I wanted to bring several people equipped with metal detectors in order to conduct my search.

The landowners agreed and for the first search, I brought only a few people along. Once we found an artefact that garnered everyone's interest, they were open to our group returning to hunt a second day. We had carefully filled in all of our excavation areas and had kept the gates to the field closed to prevent livestock from escaping. We also hauled in quite a bit of junk metal from the field. The farmers seemed pleased that we had cleared shotgun shells, pop tops, beer cans and various iron rubbish from their land.

The manner in which you conduct yourself will help determine if you are to be invited back to private land again. Show the owners what you have found on their property. Trust will be lost if they feel you are acting nervous or trying to hide what you might have found. Respect the laws in your country by offering the landowner their rightful share of your finds. Even if a percentage is not specified in your country, it will be a generous act to offer them some interesting coins or artefacts.

Nigel Ingram and other European detectorists routinely offer their services to search for any lost farm implements in the future as a means of thanking the landowners for allowing permission to detect their fields. Other detectorists offer small thank-you gifts to the landowners at holiday time as a showing of their gratitude. Such offerings often go a long way in opening doors for future hunts.

Offer to show landowners or their visitors how your metal detector operates. They often have interest in what you are doing and you might just get someone else interested in your hobby!

HUNTING TECHNIQUES: Tips from Charles Garrett

Charles Garrett has hunted with his metal detectors in England, France, Spain, Italy, Germany and Scotland, in addition to trips that have taken him from Mexico to Egypt and literally around the world.

During the past 40 years, he has written numerous articles and books on treasure hunting tips and techniques. Garrett was recently asked to share a few strategies that he used during his own trips to metal detect in Europe.

(Left) Charles Garrett and his son Vaughan Garrett, seen in the Garrett Museum, display some of the Roman and Greek coins they found together on a European metal detecting trip.

"Some of these tips might seem obvious but they are ones that I have employed while detecting in Europe," Garrett says.

- **Start by searching in a place where history is likely to be.** Treasure, of course, can be found anywhere if you are satisfied with finding only modern Euros and other recently lost items. If your goal is to find Roman, Celtic or other Old World items, then you should obviously start by hunting areas where such people once lived.

Study the history of the area where you plan to visit. Where were the oldest villages or settlements? Water sources (natural lakes and rivers) were essential to survival in ancient times. In some cases only the faintest evidence of an ancient castle's stone foundation might remain. I spent a week one time just to search the area around an early English fort site used in the late 1700s to defend against French troops. I was by myself and I literally worked from dusk to dawn to make the most of my limited time. By the end of the week I had accumulated some 500 artefacts. These included handmade lead gaming pieces, coins, musket balls and all sorts of projectiles, uniform buttons and two gold officer's cap emblems, one of them a gleaming beauty.

My preferred metal detecting method was in the All-Metal mode with my sensitivity set to detect as deep as possible. Fortunately, I did not have to fight heavily mineralised ground conditions in this area. Because of this fort's somewhat remote location, I had little tourist trash (cans, pop tops) to contend with.

- **Study the terrain in the location where you choose to hunt.** Imagine *where* coins or artefacts would most likely be lost, hidden or naturally deposited by the effects of rain, erosion and time. That is how I started my scanning of this English fort site on my first day. There had once been a moat that surrounded this fortification and an old dirt mound about six feet in height was still present. This, I decided, was the best place to start. Sure enough, I had my first signal atop this hill in little time and it was loud! The target, pinpointed at about eight inches, was a virtually pristine British Monmouthshire 57th Regiment breastplate.

- **Exercise care while recovering your treasure targets.** The story of how I recovered one of the two gold officer's cap emblems near this fort is a prime example. It was on my last day of hunting and dusk was approaching. Near the remnants of an old stone wall, I picked up a good signal and began to dig. The ground was very tough and I had to use a pick to chop through the soil. I was exhausted after more than 12 hours of hunting in the sun with only a sandwich break. In my haste to find the source of this signal before dark, I destroyed my treasure. Upon the last swing of my pick, I saw a glimmer of gold and knew that I had chopped right through it. The gold cap emblem was smashed by my retrieval pick!

I'll never forget this "haste makes waste" lesson. Since that time, I have always made it a point to use a quality pinpointer to prevent damaging a coin or artefact with my shovel.

• **Treasures can sometimes be found in less obvious places.** For example, I've learned to scan attics, basements, floorboards and even walls in old homesteads to look for hidden caches. During one of my European excursions, we obtained permission to hunt on private property that backed up to a famous 19th century battleground.

Each member of our group found musket balls this day. Noting one particularly large old tree stump, I ran my detector around the rotting old stump. Sure enough, the detector sounded off and I began to hack into the stump. We retrieved a number of smaller lead balls, believed by our group to be pistol shot. In an area of heavy shooting, it is only natural that trees will catch their fair share of flying metal.

• **Be aware of the dangers of more modern military artefacts.** Years ago, I made a hunting trip to an area near Koblenz, Germany. Some of our German friends were proud to show us various World War II German Army artefacts they had discovered with their metal detectors. We returned to one of these areas and our team recovered an estimated 2,000 pounds of artefacts. We dug countless bullets and ammunition clips as well as helmets and even hand grenades. Such military artefacts should be treated with great caution. When in doubt about discovered ordnance, notify your local authorities versus attempting to dig it up. European antiquity laws have become stringent regarding exactly how such discoveries must be reported. Always follow your country's law regarding treasure recovery.

• **Jump on an opportunity right away when it presents itself.** I can't begin to tell you how many times I've kicked myself over great treasure hunting opportunities that I've let get away from me. We found the stone wall ruins of an old stage coach stop one evening near dusk. We had no time to detect this day, but planned to return again before our trip ended. Well, it didn't happen.

People may offer to let you hunt on their property. If you wait too long to take advantage of an offer, the property might change hands or the landowner might have a change of heart. *The lesson here: don't wait!*

The caretaker of an old castle in Spain we visited offered to let me come back sometime and do some thorough searching. Knowing how things often worked out, I decided to at least scan a few minutes before we had to leave. In the end, we did not make it back there but I did make a great recovery during that short time of searching: an ancient crossbow point. For once, I was proud of myself for seizing the moment!

(Above) Charles Garrett shows some of the early coins he found in Europe. Many are from a cache found in a ploughed field.

(Above) These are a few of the interesting buttons found by Charles Garrett while metal detecting in the UK. The largest button on the top left is a 2.7 cm George III pewter coat button. The other two on the top row are from the same era and were used on a waistcoat. Bottom row: George III silver-plated livery button (left) and a 1 cm spherical button from the Elizabeth I period.

(Right) More of Charles Garrett's European coin finds. These range from early Roman bronze to more modern English coins. Moving left to right along the top row are Victoria pennies from 1887 and 1882 and an 1806 George III halfpenny.

(Below) Garrett searches for artefacts near a fort where British and French troops once clashed.

Marx Ickx of Belgium prefers to use a rugged army shovel to recover targets in this rocky soil. His efforts this day were rewarded with an early Roman artefact *(see inset)*.

CHAPTER 4

RECOVERING YOUR TREASURES

Once your metal detector indicates that a treasure target is below your searchcoil, you will want to recover it efficiently and without causing damage it. The methods for recovering targets vary as widely as search techniques. The purpose of this chapter is to present various options and recovery tools for you to test. You can then decide which best fit your particular style.

Many advanced metal detectors offer a pinpointing mode that enables the instruments to produce a signal as they hover motionless over a target. Detectors without a pinpointing function can also be used to effectively pinpoint a target with just a little more effort. In the latter case, scan back and forth over the target area to determine the spot where you receive the strongest (or loudest) signal. Experience and practice will show you how to pinpoint your target with your searchcoil.

I generally pinpoint the item and take note of the depth my detector indicates. In order to avoid damaging the item, I break through the surface of the ground with my shovel in a circle slightly larger than the pinpointed area. Then, I flip over a chunk of earth just below the indicated depth. I scan the excavated sod and often find that the target signal is now in the removed soil. If not, I sweep my searchcoil over the hole. If the target is still in the ground, I dig another scoop of soil and scan over the excavation pile to insure that the object has been moved to the surface. At this point, I turn to my Garrett *Pro-Pointer* to quickly pinpoint the

A quality pinpointer can help with the recovery of small items and also prevent damaging the target. The author points to the area his *Pro-Pointer* indicates a target to be located *(upper left image)*. When the signal is narrowed to this clump of earth *(lower images)*, the soil is carefully broken up by hand to reveal this post-medieval button.

object in the dirt pile. I always use this pinpointer to make another sweep of the recovery hole as well. You will often locate a second target by double-checking the area.

Some old-timers will swear that they have no need for a pinpointer; they can find any target with only their shovel and their searchcoil for pinpointing. While attending rallies, I have watched a number of detectorists of varying skill levels as they work to unearth their targets. Some become quite frustrated or even give up when a small object proves to be elusive. Once such detectorists are offered the chance to try out a good pinpointer, they often are left wondering how they ever lived without one!

Small coins, buttons and musket balls are often camouflaged well by the soil colour in which they are buried. The use of a handheld pinpointer can greatly speed your recovery time of such items and help prevent damaging them.

Take proper care to dig beyond the pinpointed area of your target. Nigel Ingram of England damaged this bronze bell, an ornamental piece from the 1700s era, when his trowel struck it during recovery.

Treasure Recovery Tools

The implements used by detectorists to unearth their treasures vary based upon soil conditions. Small **hand-held trowels** can be used to dig coins in softer soil in the city or in your yard—areas where you want to minimize the chance of damaging a target. Some detectorists prefer to **a dull point screwdriver or a treasure probe** to pop coin targets from the earth. After pinpointing, push the probe into the ground three or four centimeters behind the coin, sticking the probe in at a 45-degree angle about seven centimeters deep.

With your treasure probe inserted to this depth, push forward and to the left, making a slit in the ground three to five inches long. Then make the same slit to the right, with the slits leaving a "V-shaped" piece of sod which you lift and push forward, swinging it up and out of the ground.

After you have retrieved your coin, the sod will fall back into the hole in the exact place it came out, and the grass roots will not die. This is especially true if you don't cut roots of the grass

This group of hand-held treasure recovery tools shows the wide variety of instruments that metal detectorists prefer. The razor-edged digger at left is ideal for sawing through roots. The two longer, wooden handle digging tools have blade edges to chop through roots and teeth to rake through soil. Also seen are polymer and steel digging trowels and a treasure probe *(fourth from left)* used by many coin hunters.

by making your "V," but merely force most of the roots from the ground.

European detectorists who hunt in the pastures or in the forest must use more rugged **spades or shovels** for their target recovery. Suitable, compact spades can be purchased at many hardware or gardening supply stores. There are also a number of retailers who sell digging tools specifically designed for the needs of the serious treasure hunter. You can locate such suppliers by scanning treasure hunting magazine or with an Internet search. Read reviews about various recovery tools on detecting forums to see which tools are favored by detectorists. Your own trial and error is, of course, the best way to discover the tools that work best for you.

Beach hunters can use their same favored shovel for digging in wet and dry sand, but often they will use treasure tools better-suited for the water. A rugged **plastic sifting scoop** works well

(Upper left) Many surf hunters rely on a long-handled galvanized metal recovery scoop. *(Upper right)* This battle artefact hunter prefers a straight-handled steel spade with foot pegs above the shovel blade for powering through tough roots. *(Below)* A European detecting team submitted this photo of their recoveries, their detectors and their recovery tools (including pinpointers and a small, folding camp shovel).

In this photo series, a beach hunter recovers a gold necklace from the surf using a long-handled galvanized metal scoop.
(Upper left) Bearing down with his foot, he scoops under the target and up through it. *(Upper right)* He then sifts through the sand and repeats the process if the target is not found.
(Right) Mastering this technique can result in terrific discoveries, such as this valuable necklace.

in dry sand to recover coins and jewelry. In the hard-packed wet sand at the water's edge or in the surf, you should invest in a sturdy **metal sand scoop**. In the surf, such metal scoops are often mounted on steel poles to allow the hunter to retrieve underwater targets without being knocked over by waves.

In addition to digging tools, you should wear a treasure pouch, coin-hunting apron or some other special bag in which you can

Stainless steel sand scoops *(left)* are best for surf and wet sand recoveries, while a rugged plastic sand scoop *(right)* works well for dry sand recoveries.

collect both your treasures and your trash. Use a zippered pouch or compartment to best protect against the accidental loss of your recovered treasures while you are in the field. A good treasure apron should have at least two pockets made of either waterproof plastic or with plastic pocket liners. Quite often you will be digging in areas that are wet. The damp soil accumulated on recovered targets can cause the contents of non-waterproofed pockets to leak through onto your clothing. You can also protect your clothing (and your knees) with comfortable and adjustable waterproof knee pads.

Treasure Recovery Techniques

Your methods for excavating targets will vary based upon your location. In areas where you want to minimize damage to grass and its roots, you will use a more careful technique. You are obviously less concerned, however, with how you a dig a target hole in a ploughed field, in the forest or on the beach.

The beginner will often dig larger and deeper holes than necessary while recovering a treasure target. He or she will also often struggle to find their smaller targets. Here are a few suggestions to help improve your target recovery efforts:

• *Dig only slightly below the depth you have pinpointed the target.* Many modern metal detectors with target identification also give a target depth indication. So, what do you do if you are using a detector that does not or cannot indicate true target depth?

First, precisely pinpoint the target as you normally would. Then "size" your target by slowly lifting your coil above the ground to gauge its relative size and depth. If you are coin hunting and have a nice silver response you believe to be a coin, you will loose this signal as you gradually lift the searchcoil higher. If you continue to get a strong response even after lifting the coil about 15 to 20 centimeters off the ground, you're certainly dealing with a target much larger than a coin. By practicing this "coil lift" method in your test patch at home on targets that you have planted (where you know their exact depth) you will be able to reasonably judge your target's depth.

• Check the target hole and the excavated soil to determine whether you have moved your target. *If you have lost all target response, chances are that your item may have dropped down deeper into the hole you've dug.* If you don't have a pinpointer to probe the hole, remove more soil and swing your coil over it again.

• If you don't find the target in the additional dirt you've dug, *the target may be on its side in the large original chunk of earth you removed.* Spread out this soil with your foot or shovel and sweep it with your coil. If you still can't find the item and have been searching with discrimination, switch over to All Metal Mode. Your item may have slipped against an iron object that is now masking the item's presence.

• *Be aware that the hole you've created can cause false signals if you are searching with threshold tone.* Your detector has been ground balanced to the soil you are searching and it is now suddenly facing a void of this soil content.

Once you have recovered a fair number of treasure targets, all of the above information will be second nature to you. The next consideration is how to actually dig for your target. Listed below are a few common methods for recovering your treasure based on your location and available treasure tools:

- **Plug method:** Make a plug in the ground by pushing your spade down around the area that you have pinpointed with your detector. After punching around in a circle or square, flip the plug over and scan it to see if your target is inside the plug. If the plug offers no response, scan over the hole and the ground around the hole. Usually, if you have dug the plug hole to the depth that the target was indicated, the target should be easy to recover. After retrieving your treasure, carefully push the loose soil back into the hole and replace the plug.
- **Loose soil method:** In areas where digging a neat, clean plug is not necessary or practical (such as on ploughed farmland or in the forest), simply dig the amount of earth needed to reach the target depth. Scan the removed dirt and the hole. If your detector indicates that the target response is still in the hole, dig deeper and repeat the process. Once you can get a target response from your pile of dirt, push it around with your foot or shovel and then scan over the pile again until you can locate the target. Keep pushing the dirt and rescanning it until your target is spotted. Experienced detectorists can speed through this entire sequence in mere seconds for the majority of their recoveries.
- **Small object recovery:** Sometimes, you might dig up a very small target that is nearly impossible to spot with one of the above methods. In the absence of a hand-held pinpointer, use your detector's searchcoil as your pinpointer. Once you've narrowed down the area where the target remains, scoop handfuls of dirt and pass them directly in front of your searchcoil about 2 to 3 cm away. (Make sure that you are not wearing any metal objects such as a ring or watch on your hand passing the dirt in front of the coil.) Continue scooping and inspecting dirt until your detector responds. Then, carefully sift through the dirt in your hand until the item is located. If you sift through the soil until nothing remains, you've obviously dropped the little item and must start again.

This veteran detectorist searches for a tiny treasure target using his searchcoil to help locate the item. Note that as he passes handfuls of dirt across his searchcoil that he is wearing no rings or other metallic items on this hand.

Treasure Recovery with a Pinpointer

Although hand-held pinpointers have been around for years, many metal detectorists have not learned their true value. Some people have not tried hunting with a pinpointer while others have decided against spending the extra money. Those who have used

(Above) The author relied upon his *Pro-Pointer* to pinpoint these tiny European items. Left to right are a Roman nummi coin (circa 300–400 AD), a 17th century hammered coin found in Italy and various buttons recovered in England. Each target is quite small in size and blended in well with the excavated soil.

a well-designed treasure pinpointer will often tell you they would not go into the field again without one.

Even veteran hunters will be challenged at times to find small targets. The ploughed fields often hunted by Europeans are a prime example. During the rallies we attended, I watched quite a number of people sifting and scanning the clods of earth to find their targets. Dry farm soil falls back into the hole almost as quickly as you dig it. Below the dry soil is often a clay layer which comes up in dark clumps that are hard to pry apart. Small coins are often very difficult to pick up even when your searchcoil indicates that they are out of the excavation hole and in the surface soil pile.

A good pinpointer is indispensable to:

- **Speed target recovery.** The high-quality *Pro-Pointer* from Garrett has side scanning ability and a scraping blade which allows you to rapidly sift through a large concentration of soil to home in on your treasure.
- **Accurately locate small objects.** Some targets, such as tiny hammered Roman coins, simply blend in with the soil and are time consuming to dig out with your fingers. Darker coins often blend right in with the soil colour, further complicating your efforts to find what caused your detector to respond. Your pinpointer will generally lock in on a tiny metallic item faster than you can spot it with your eyes.
- **Prevent digging large holes.** In most public areas, you will want to dig the smallest recovery hole possible. A pinpointer helps eliminate useless extra digging by zeroing in on your target.

A *Pro-Pointer* helped find a second artefact in this recovery photo. The archaeological team had successfully recovered a musket ball from this deep hole (about 32 cm). A final sweep of the hole by the author revealed another target in the side wall. Careful excavation revealed a second musket ball *(below arrow)*, seen still embedded in the side wall.

• **Identify multiple targets in close proximity.** Even after you have recovered your treasure target, it is wise to use your pinpointer to make a final sweep through the soil you have uncovered and to also make another scan of the excavation area. You might be surprised at how often your pinpointer will announce the presence of secondary or smaller targets that you were not expecting.

• **Search tight interior areas such as walls and ceilings.** Caches of coins and jewelry hidden away within an old homestead can be difficult to detect with a standard metal detector because of the proximity of pipes and other metallic objects. A pinpointer can reach into tight wall and floor spaces for concealed hoards. It will also locate nails, metal wall studs, rebar, conduit and control boxes that are out of sight behind sheetrock or flooring.

In loose sand and soil, sift or scrape through the excavation pile, flattening it out as you push through and over it with your pinpointer. For rugged surfaces—those full of rocks, roots, and other debris—Franco Berlingieri of Belgium suggests a "quick-find" technique that he has developed with his Garrett *Pro-Pointer*.

(Above) Detectorist Dave Totzky sifts through soil with a pinpointer on a battleground. He was able to locate a complete buck and ball load *(seen at right in his glove)*, consisting of a larger, slightly-mushroomed musket ball and two single-ought pieces of buckshot. His pinpointer helped him find the two smaller, and unexpected, artefact targets.

Franco places his *Pro-Pointer* flat on the ground above the spot where his detector indicates the target to be. Then, he twists the pinpointer in a clockwise circle until it begins vibrating and audibly alarming. Where the audio and vibrate alerts become the strongest, he stands the *Pro-Pointer* up to use its pinpointing tip to precisely locate the small treasure object.

Your treasure recovery methods and the tools you choose to use will vary based upon the terrains you encounter. Just as your ability to locate treasure with your metal detector improves with practice, your ability to quickly recover targets without damaging improves with repetition.

The final step in target recovery is the need to *fill in your excavation hole.* Do not leave unsightly and potentially dangerous holes in the property where you hunt. You should always leave your hunting ground in better condition than you found it. Bring a treasure pouch or bag to collect the garbage metal that you dig. The land owners will be more apt to allow you to metal detect on their property again if they see the pop tops, tin cans and other debris you haul away.

Bob W. of England used his *Pro-Pointer* to pinpoint this Charles half groat, circa 1526 to 1544 AD. It has a value of about £225.

Chapter 5

European Coin Shooting

Metal detectorists often take on interesting sobriquets related to their particular detecting interests. Gold hunters call themselves prospectors. Coin hunters often call themselves "coin shooters" and say that they are going coin shooting.

I happened to meet a couple of pretty efficient coin shooters while in England, Gary Norman and Jason Price from the Leicester area. These friends both use *ACE 250* detectors with great success. I spent some time with both of them in Field 7, an area which had been producing Roman artefacts at the Oxford-area rally we were attending. Gary found two Roman coins during his first day of hunting. They tried their luck in other fields for much of the following day but were compelled to return to their lucky Field 7 by the afternoon. Their good luck held and they found several more Roman coins.

Gary and Jason regularly hunt on 150 acres in the Leicester area with the permission of the landowner. "Before we go out to hunt, we Google the history of the area we will be visiting to get a feel for it," Gary explained. This land proved to be productive for them. "We were walking past a treeline and my *ACE* signaled a good target. It registered on the Target ID by the five-cent piece icon, which we know normally will be a good hammered coin."

Jason pinpointed this signal and dug up an Edward I farthing from the period of 1297–1302 AD. Within 15 feet of this discovery, Gary found a silver Roman denarius from the reign of Marcus

Coin-Hunting Companions Work the Fields

(Left) Jason Price, far left, and Gary Norman examine an early coin Jason has just located during a 2009 UK rally.

(Below, left and center) Gary displays two of his second day finds at the rally, a crotal bell and an 1806 George III coin. *(Below right)* Two of the Roman coins Gary found during the first day of the UK rally.

(Left and inset) Jason with a musket ball pinpointed with a Garrett *Pro-Pointer*. *(Below)* Gary's silver denarius, circa 161–180 AD. Bottom row items found by Jason are: 1883 gold sovereign *(left)*, Edward I farthing *(center)* and an Iron Age burial bead *(right)*.

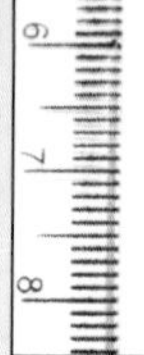

(Left) Coin shooter Jason Price found this Edwardian era 15-carat gold and diamond ring on one of his first hunts in England with an *ACE 250.*

Aurelius, circa 161–180 AD, and a Henry III hammered coin, circa 1216–1272 AD. Jason, not to be outdone, next found a 22-carat gold 1883 Queen Victoria half sovereign. "My *ACE* pinpointed this so perfectly that I found this stunning coin in the first shoveful of dirt I dug," he said.

They have also located nice jewelry pieces while sweeping for coins. "I dug a 15-carat Edwardian gold and diamond ring from 1910 on one of my first hunts with my *ACE 250,*" said Jason. "It gave the faintest of signals and registered near the five-cent icon. I dug it from claggy mud near a little stream."

"What we enjoy about coin hunting in England is that we can find early British coins as well as Roman, Medieval and Iron Age items on the same land," said Jason. For example, Gary unearthed the leg of Medieval bronze cooking pot on one of their hunts. Nearby, Jason received an "absolutely banging signal" from his detector. "It was a burial bead made of bronze that had nice patina coating. Our local FLO identified it as being from the Iron Age. What impressed me most about this find is how accurate my *ACE* is for pinpointing small items. It was dusk and I had to dig a long way down in near dark conditions. Once I had the bead out of the hole, I pinpointed again and literally put my hand right on it."

The Changing Faces of Currency

Staters, groats, skillings, guilders, sovereigns, guineas, francs, thalers, marks, kopecks, balboas, Euros, reales…the names of the world's coins are as varied as their designs and compositions.

(Above) George Clerinx of Belgium holds a golden medallion, circa 1847, found with his *GTI 2500* in a field near the ancient Roman town of Tongeran, the oldest Roman settlement in Belgium. The two photos at left show the obverse and reverse of George's medallion.

(Below) Other French coins, circa 1830s, found by George with his *GTI 2500*.

This image shows the obverse and reverse of a Spanish "Pillar Dollar"—an 8-Real recovered by Robert Marx from the *Rooswijk*, which sunk in 1740 off the southeast coast of England with 360,000 guilders in gold and silver species. The third coin image above shows how Spanish reales would have been cut into eight bits, or "pieces of eight."

Regardless of what country you visit, coin hunting with a metal detector can be productive *almost anywhere*. Coin hunters in the United States are overjoyed to find a coin that dates back to the 1700s. In Europe, however, it is common for treasure hunters to dig up currency that predates the time Christ was on this earth.

The world's first coins were minted in western Asia Minor in ancient Lydia (modern-day Turkey) some 700 years before the birth of Christ. They were made of a metal known as electrum, a naturally-occuring alloy of silver and gold. The Lydians melted the electrum into circular shapes and stamped the sign of their king into them.

These coins were followed around 600 BC by the first Greek coins, which were stamped with designs on their obverse (front) side. Various designs were then hammered into the coins that began to be created throughout the Mediterranean cities during the ensuing centuries. The early method of coin-making involved the artisans placing a blank piece of metal between two iron dies. The top die was then struck with a hammer to produce an image on both sides of the metal. By the end of the Sixth Century, most Greek punches had a die for the reverse side of the coin as well. The first silver coins were minted on the island of Aegina and were stamped with a sea turtle motif. The Greek world was eventually

One of the smallest European coin targets is this "cut quarter," one-fourth of a hammered coin that was once snipped away to make change. It is seen between a 1-Euro coin and a £1 coin to indicate its scale.

divided into at least a hundred self-governing cities and towns, most of which began issuing their own coins.

The Celts began minting coins around the Fourth century BC and continued to do so for some 400 years. Celtic coins were either struck with an iron or bronze die or were cast by pouring molten metal into a set of molds which were broken apart after the coins had cooled. The markings of Celtic coins—often depicting horses, gods or warriors charging into battle—were influenced by ancient Greek coins. Cast Celtic coins from around the First century BC are found on occasion today in Great Britain, France, Belgium, southern parts of The Netherlands and other areas. Such coins are easy to spot because of their thick, bowl shape. The Celts often cast such coins from bronze, copper or silver. Others were made of electrum, gold or electro plate.

The earliest known Roman coins were lumps of raw bronze (called *Aes Rude*) and other base metals which were eventually replaced by cast pieces of bronze. The first true cast Roman coin, the *Aes Signatum* (signed bronze), replaced the Aes Rude around the start of the Third century BC. Each Aes Signatum was cast at a weight standard of 1600 grams to eliminate the need for weighing coins during merchandise transactions. This coin's hefty weight and single denomination led to yet another cast coin around 269 BC called the *Aes Grave* (heavy bronze). The ancient Romans found this coin to be more functional, as it was produced in several denominations, and its coinage became the primary issue in Rome for decades.

The first Roman silver coin was the *denarius*, which was first struck about 211 BC and continued through the middle of the Third century AD as the new backbone of the economy. The silver content and the accompanying value of the denarius slowly decreased over centuries of use. A more valuable gold coin called the *aureus* was also issued for large payment needs.

Roman coins discovered today can be recognized because they are thicker than more modern European coins and because they are generally not perfectly round. Due to the vast area of the Roman Empire, coins can be found in all of Europe to the west of the Rhine and south of the Danube. Roman coins also found their way into the areas beyond the Roman Empire, such as Poland, Russia, northern Germany and Scandinavia.

During the early Middle Ages, the Anglo-Saxons, Visigoths, Franks and Merovingians minted gold coins known as *thrymsas*. The early thrymsas were heavily influenced by Roman design but gradually developed their own patterns. As gold disappeared from use in these areas at the end of the Seventh century, silver coins became widely minted.

During the late Middle Ages (476–1250 AD), minting licenses in much of Europe were given to the cities or to dukes, bishops and counts in the 12th and 13th century. The silver penny became common and after 1250 AD, silver groats, English shillings and southern Dutch shillings were in production. Spain acquired large amounts of South African silver and began producing thicker silver coins than those turned out by other countries. Gold was also used in this period in such mintings as the French shield and the gold nobles.

By the 17th century, the world's *hammered coins* were being replaced by *milled or pressed coins* fashioned by machines. Another early coining method of pouring molten metal into a cast or mold created what are known as *cast coins*. More often than not, cast

The thrill of a good recovery: Mark Hallai and Franco Berlingieri celebrate the recovery of a Napoleon coin found near an old French battle site in Belgium. *(Below)* Close-up of this old Napoleon coin.

(Above) In France, newer coins found by detectorists can have higher silver content than older coins. The older, 16th-century French coin at left was made with about 20 to 30% silver and about 70% or more copper. In contrast to this composition, the more modern coin at right, an 1863 French douzain, was minted with about 90% silver and 10% copper. The earlier French coins were made with less silver because it was more scarce until the Spanish began bringing in large quantities of silver.

coins of ancient China, Greece and Rome were made of copper or bronze, although some early cast coins were made from silver and occasionally gold.

The Spanish milled dollar was a minted silver coin worth eight reales which was first created in 1497. The Spanish dollar became the first world currency and was accepted in the United States as legal tender through the 1850s. This milled dollar was often cut into pieces to make change. Each of the eight *bits* (*pieces of eight*) was worth 12.5 cents—thus the phrase "two bits" was later coined to refer to a quarter dollar.

Mechanically pressed coins began appearing in England and France during the 17th century. The resulting coins, made from cast and sheet metal, are of even thickness and equal size as opposed to the earlier coins that had been cut by hand and manually hammered. The mechanical press also turned out coins with images that are perfectly centered. In general, European coins that appear to be evenly punched or hammered date only date back to about 1650.

Another indicator of a coin's age is the lettering that is stamped or hammered into the coin's design. The oldest Roman and European medieval coins had abbreviated Latin words on their outer rim. The Latin was gradually replaced by the countries' native languages in the 18th and 19th centuries.

Today, 27 European countries have been accepted into the EU and have replaced their national currencies with the Euro. In the first two year of minting Euro coins in 1999 and 2000, mints in Belgium, Finland, France, the Netherlands and Spain produced more than 15.5 b illion Euro coins to meet the needs of the original countries.

Each of the eight Euro coins—ranging from one cent to two Euros—has a common reverse side portraying Europe and the coin's value. Each country in the eurozone has the right to create a regional obverse design, thus increasing the total number of Euro coins that can be collected. In addition, three European microstates that use the Euro as their currency—Monaco, San Marino and the Vatican City—are allowed to mint their coins with their own designs on the obverse side. These three coin designs are coveted by coin collectors and are rare to find in general circulation.

These Roman coins *(seen obverse and reverse)* were found by a Garrett *GTP 1350* detectorist in France. These coins were found on an old Roman village site at depths ranging from 10 to 20 centimeters by a group of 25 European detectorists who found a total of 80 silver coins and eight gold coins in one day. The silver coins hold values of about €500 each while the best of the gold coins is worth €50,000. This particular gold coin is valued at about €20,000. The first Roman coin found that day was recovered by an *ACE 250* user.

European Coin Hunting Tips

The use of a high quality metal detector is very important when searching in European countries. Some of the ancient hammered coins are especially thin and small in size. Such coins have often settled deep in the soil over the many centuries since they were lost. The value of old European coins varies widely based upon their age and condition. Generally speaking, the smoother a coin feels after it's cleaned, the more it has been in circulation. The best coins you find still have a rough texture where you can feel the engraving or stamping. Such rare virgin, uncirculated coins have more value.

Many hobbyists seek permission from local farm owners to search their fields during times when crops are not in season. One of the best times is right after the first ploughing of the fields each spring. Farmers often turn up the land and bring new artefacts toward the surface. It is not uncommon to find centuries-old Roman coins and artefacts. See chapter 15 for laws in European countries regarding what items may be kept and what must be reported upon discovery.

Many of the fields in Europe have rich history and have been worked in some cases by farmers for thousands of years. The quantity of good finds that you make is directly proportional to the amount of time and effort you put into your searches. Here are a few tips to help improve your success rates:

- *Be methodical in your search area.* As you walk an old homestead or field, establish a pattern for thoroughly covering the area. Even a previously worked-over area can be productive if you give it due diligence. Be sure to overlap your sweeps as you advance your searchcoil. One suggestion is to walk a zone back and forth with your searchcoil overlapping the previous area by about one-third to one-half of its area. Once you have found a nice coin or artefact, grid out your search area and work it carefully to see if other items exist in this spot.

• *Think about where a natural resting spot might have once been.* Franco Berlingieri found a group of 20 gold coins near a large, round boulder in the woods that may have once been near a farmland path. The impressive collection of gold coins—mainly from the 1800s—found beside an old tobacco smoking pipe, were likely once contained in a leather coin purse. Franco believes a traveler may have sat down by the stone to rest and smoke, fell asleep, and lost his coin purse after he awoke.

• *Look for broken pottery, lead or other evidence that people had once lived or concentrated in a particular area.* Farmers often plough up bits of old pottery. These pieces can often be spotted on the surface as you sweep the fields. Areas where significant shards of pottery have surfaced are often very productive areas to find other metallic artefacts.

• *Be careful with your discrimination settings.* Some of the best ancient hammered coins are very thin and often produce only a very faint signal. Discrimination settings that eliminate iron, pull tabs or aluminum may also cause you to miss such coveted small coin targets.

• *Headphones are highly recommended* to help block out external noise and to enable you to pick up the faintest signals of those really deep treasures.

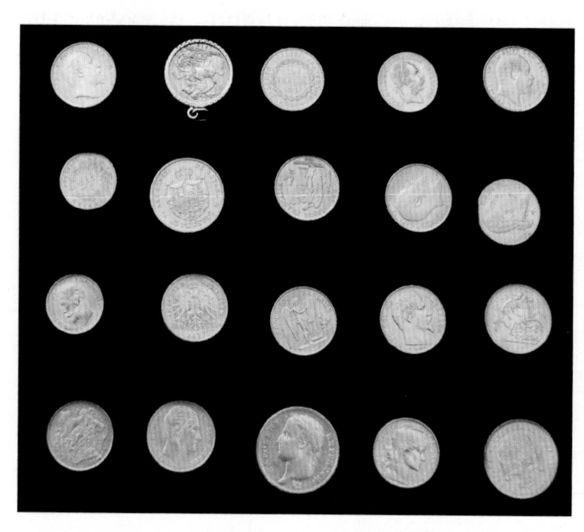

This group of gold coins, dating from 1802 through the 1930s, was found by Franco Berlingieri in Belgium near a traveler's resting spot.

Belgian's Favorite Find Was "Five Minutes from my Home"

Marc Ickx *(left)* found this 1619 hammered Albertus and Elisabeth silver coin *(above/below)* a short distance from where he lives.

You don't have to travel far in Europe to find nice coins. Marc Ickx of Belgium bought a Garrett *ACE 300* in 2002 and joined "this wonderful hobby. Very soon I started to find some coins, buttons, and lots of other stuff so I knew that this was gonna be my hobby for life," he related. "I'll never forget the moment that I found my first Roman coin—a very beautiful sestertius of Faustina's mother."

Marc was so thrilled with his new hobby that he sold his first detector after a year and bought a *GTI 2500*. "Over the years I did find some very nice things but this coin is my best find yet! It's made of pure silver with a cross cut of 5 cm dated 1619 and in very good condition," Marc noted. "Needless to say, my heart stopped beating for a few seconds when I dug it up. I found it in a garden of an old house only five minutes from my home."

• *Keep an accurate record of the numbers and locations of coins and objects you discover.* Your journal may point out trends regarding certain areas that could benefit your future hunts.

• *The value of researching areas for potential hot spots is always a key to success.* Talk to older residents in the areas where you plan to hunt. Some of them may share interesting stories—such as where their grandparents ran a country market—that could lead you to significant recoveries which can have a positive impact on your budget. In 1984, Leo Kooistra of the Netherlands found an old

Peter the Great silver coins are so tiny (smaller than 1 cm) that Russian detectorists often refer to them as "fish-scale" coins.

cache of 68 silver coins that dated to about 1,000 AD, a prized find that he sold for 12,500 guilders.

Some European coins are extremely tiny and are quite challenging to find. Sergey Chernokryluk of Moscow and his detecting companions search local farmlands after the farmers have ploughed their fields. Sergey has found many 5-kopeck coins dating back as early as 1796 but his favorite finds have been tiny silver Peter the Great coins. Peter the Great (officially Peter I) ruled from 1689 to 1725 AD and was the last Russian czar to strike coins in the silver wire money form. They often are very deep and do not register as silver on the detector's Target ID scale. "They register nearer to fine iron," Sergey says. These thin coins are known as "fish-scale" coins because of their small size; some are about 8 mm in size and can weigh .28 grams.

Sergey hunts with a Garrett *GTP 1350* metal detector and generally finds at least several old coins on each hunt. "I recommend using a Double-D coil with an all-metal or non-discrimination program to find these small coins. It is necessary to dig any faint signal because it might be one of these Peter the Great coins."

CAESAR COINS OF THE ROMAN EMPIRE

JULIUS CEASAR
circa 44 BC
Silver denarius

AUGUSTUS
AD 27 BC—14
Silver cistophoric tetradrachm

TIBERIUS
AD 14—37
Gold aureus

CALIGULA
AD 37—41
Brass sestertius

CLAUDIUS
AD 41—54
Silver cistophoric tetradrachm

NERO
AD 54—68
Brass sestertius

GALBA
AD 68—69
Gold aureus

OTHO
AD 69
Gold aureus

VITELLIUS
AD 69
Brass dupondius

VESPASIAN
69—79 AD
Copper as

TITUS
79—81 AD
Brass sestertius

DOMITIAN
81—96 AD
Silver denarius

GALLERY OF EUROPEAN COINS FOUND WITH GARRETT METAL DETECTORS

(Right) This 1807 Czech Republic hammered bronze coin (4–5 cm diameter) was found in the mid-1990s near an old castle with a Garrett *GTA 2000.*

(Below) These are four of UK detectorist Nigel Ingram's favorite coin finds. Left to right are a Roman silver coin, a 1400s-era British gold noble coin which was bent by a farmer's plough, a 1587 medieval Elizabeth I silver sixpence, and a 1775 George III gold guinea coin.

(Above) These tiny silver Greek coins, found in Sicily, each weigh about .8 grams and measure only .8 cm in diameter. They were hammered from bronze dies around 500 BC.

(Above) This group of old hammered, irregular-shaped Roman silver coins was dug since 2000 in Belgium. They include images of King Philip and date from 280 to 300 AD, each measuring from 2 to 2.5 cm in diameter.

(Above) The obverse of this Greek silver coin, circa about 500 BC, shows the face of Alexander the Great. The reverse shows an eagle perched upon a branch. This small hammered coin is about 1.5 cm in diameter and weighs 7 to 8 grams.

These two Greek coins were cast in silver between 463 and 411 BC. They are 3 cm in diameter, weigh about 16 grams and were also found in Sicily.

The nearer Roman coin is named Light Miliarensius and the one to far right is named Gloria Romanorum. Both are very rare Roman coins circa 379 to 395 AD, and are worth about €4,000 today.

Franco Berlingieri found these two coins one meter apart. At left is the Juliatiti (80 AD) and at right is the Marciana (114 AD). Both are hammered silver valued at about €3,500 but Franco considered them special because the Juliatiti has his daughter Tiziana's nickname "Titi" in its inscription.

Hammered Greek coins such as this one have great value. This 2 cm diameter coin was found in Sicily and is virgin (uncirculated). The obverse shows Alexander III and has a winged lady on the reverse. It was originally created between 323 to 336 BC.

(Right) This bronze Roman hammered coin shows the ruler Hadrian on the obverse. It measures 3.25 cm, has an approximate value of €1,000 and was circulated between 117 and 138 AD.

(Right) Although more modern, this Italian coin was an interesting find for one coin hunter because of its rich history. The obverse depicts Mussolini from 1945 and the reverse includes an inscribed quote of Mussolini: "Better that I live one day like a lion than 100 days like a sheep."

(Above) This group of gold coins found in Holland with Garrett detectors range in dates from 1587 (top of grouping) through 1879 (bottom right 10-guilder King Willhelm coin). One gold guilder was worth ten silver guilders in the early Dutch currency.

(Right) Early coins were often reused after a change in the empire command brought a new ruler. This 2.5 cm Roman coin was from the Augustus era (27 BC to 14 AD) but it was restamped by someone in later years to reuse the coin. The reverse side depicts the Altar of Lugdunum from the Augustus era.

(Left) These tiny Greek staters were hammered in the city of Obol around 6 BC. They are 1/24 of a full stater, cut down from a full size coin. Their obverses show a scorpion and a turtle, while their reverses show a stamp of the empire.

(Above) Three rare Roman hammered gold coins. At left is a Maximum I from 235 AD, valued at €8,000. In the center is a Faustina Junior from 145 AD, valued at €1,600. At right is a Vespasian from 69 AD, valued at €2,750. These types of coins *(their reverse sides shown below)* were struck between 323 BC to 69 AD.

EUROPEAN CAST (MOLDED) COINS

(Right) These bronze Celtic coins were cast by the Ambiane people some 2,000 to 3,000 years ago. They are among many that have been found in France and Belgium with Garrett detectors. Each of these coins are about 1 cm in diameter and weigh 2 to 3 grams.

(Left) This group of molded Greek coins was found in Sicily on a beach near Thasos. They are made of silver, weigh about 10 to 12 grams each and are 2 cm in diameter.

(Below) Early Greek silver coins were cast after a piece of silver had been cut from a larger bar such as this unfinished piece of silver found with a Garrett metal detector by Mark Hallai of Belgium.

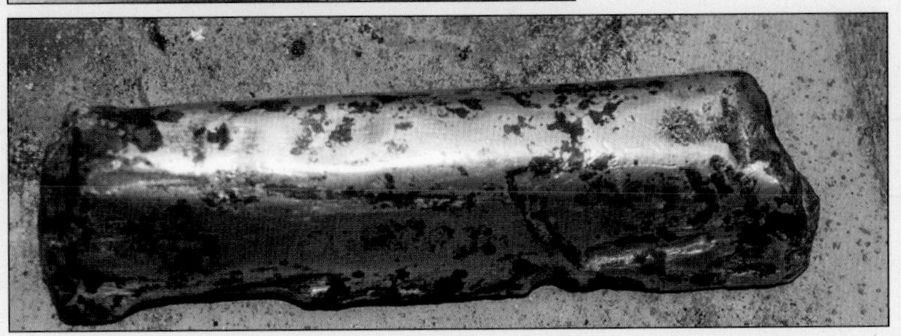

(Left) This Celtic coin dates back to around 500 BC and is only 1.5 cm in diameter, weighing 4.2 grams.

(Above) Two small silver Celtic coins, circa 100 to 50 AD, found in a farmer's field near Lyon, France. *(Below)* A 2nd Century bronze Roman denarius coin known as an Antoninus Pius, which was struck after the ruler was already deceased. Born in 86 AD, Antoninus Pius was regarded by many as one of the "five good Emperors" of Rome. Many different coin designs were hammered in his honor into both silver and bronze between 138 and 162 AD. The obverse of this denarius shows a bust of the Emperor facing right. The reverse depicts a square altar with double doors.

CHAPTER 6

HISTORIC HOARDS

Bruno Lallin of Paris knows that many great treasures can be found hidden inside older European homes. In fact, he has found quite a few such caches. He was contacted by one family to help recover valuables after they found paperwork documenting gold coins and gold bars once owned by a relative. The family, aware that their loved one had *not* cashed in this gold prior to his death, assumed that he must have hidden the valuables somewhere in or around his homestead. Bruno agreed with the family that he would give them 70% of whatever he found and he would keep 30% as payment for his detecting services.

Bruno used a Garrett *GTI 2500* to search the property around the house and the walls and floors within the old home. His search was soon concentrated in an earthen basement located at the back of the home. Bruno found the ground to be highly mineralised and the basement tricky to negotiate with his detector because of the presence of many old wine vats with metallic bands. His efforts paid off, however, when his *GTI 2500* announced the presence of a deeply buried metallic object in the basement.

Bruno dug to 32 centimeters depth and unearthed a metal box that contained 210 gold coq coins. The largest of the coins was a 20 franc French gold coin from 1910 that is valued at about 700 €. Bruno happily claimed his 63 gold coins in payment but to date the gold bars mentioned in the family's paperwork have yet to be discovered.

Not all great caches are found in the farmer's field. Take for instance this pile of Napoleon silver recovered from the inner walls of an old house. The searcher found more than 200 French coins in this cache, the most valuable of which he sold for 300 €.

(Left) Bruno Lallin and his detecting companion use an extra large coil to detect deep beneath the attic flooring in this old castle.

(Below, right) Targets that were detected required tearing out the flooring.

(Below, left) These are some of the 63 gold coins from the late 1800s and early 1900s that Bruno received as payment for recovering a cache with his *GTI 2500*.

The ultimate goal for many European detectorists, whether they will readily admit it or not, is to discover a hoard of ancient coins that were cached away hundreds or even thousands of years ago. Such discoveries are made fairly often, and can be seen in European treasure hunting magazines and on hunt club forums on the Internet.

The most recent buzz on the Internet concerning metal detectors is over a large haul of Anglo-Saxon treasure found in England. Detectorist Terry Herbert was exploring a friend's farm near Staffordshire in July 2009 when he made the fabulous discovery. His haul was not announced to the public for three months while the British Museum worked to insure that all of the valuable pieces had been excavated. In the end, Herbert's hoard amounted to more than five kg of gold, 2.5 kg of silver, 1.6 kg of precious metals and other copper alloy materials.

The Staffordshire Hoard, as it is now called, is an amazing collection of Anglo-Saxon war gear and other loot likely collected during wars in Europe during the seventh century AD. Among the 1,500 gold and silver items were coins, sword fittings, Christian crosses and precious metal stones. "This is what metal detectorists dream of, finding stuff like this," Terry Herbert told the UK press. The final value of this Saxon hoard has not been fully assessed by the British Museum, but the 55-year-old detectorist's share of the reward is believed to fetch at least £1 million.

Some of the grandest recoveries are hoards of silver or bronze coins hidden away by Celts, Saxons and Romans. People have concealed their valuables since the dawn of time. They buried their money, jewelry, weapons and even food at times to prevent these possessions from falling into the hands of others, particularly during times of warfare or civil unrest in their area. The conquered were often unable to ever return to their homelands—if they survived at all—and thus their hoards of goodies remain in the ground to be found to this very day.

Leo Kooistra of the Netherlands *(above)* holds coins from the hoard of 1,327 coins that he recovered using his *GTI 2500* several years ago. *(Below)* Closer view of many of the coins from this impressive cache.

Although such Celtic and Roman hoards are the treasures of dreams for many a metal detectorist, the most common caches found are generally smaller. These can include coin wallets or small pottery containers stuffed with someone's meager possessions. The larger and older hoards were often packed into large iron or brass pots and buried deep. Regardless of the cache size, it is recommended that you use a large coil to search deep for these bigger treasure targets. Such deep hoards, once out of the range of the earliest metal detectors, can be picked up by the more powerful modern instruments and such attachments as the two-box searchcoil, or Depth Multiplier.

Veteran detectorists are sometimes called upon by families to help locate the stashes believed hidden by a recently departed family member. For example, Leo Kooistra of the Netherlands recalled his biggest hoard of coins, found in August 2001 with a Garrett *GTI 2500*, was the result of a family's knowledge. He was contacted by the children of an older farmer who had recently passed away. They knew that their father had collected and cached many older silver and gold coins, some his own and some passed down to him by his father.

The children offered Leo half of what he could find if he was able to recover their father's hidden money. "I started searching their farm with my *GTI 2500* with the Eagle Eye two-box attachment," he recalled. After searching without luck for 15 minutes, Leo removed the *TreasureHound Deepseeker* attachment and tried a new tactic. "I set my *2500* to the Custom Mode and then notched out everything. I knew from experience that even without any notches present, my detector would still report a very large target. I walked through the family garden and swung the big coil around near the windows of the old house. A short time later, I got a very loud bell tone response and I knew it was bingo!"

Leo began digging and at one meter depth he unearthed a large bucket container known as an emmar. It smelled terribly, as it had been packed with old papers which had begun rotting from the groundwater that seeped into it. As he dug his hands into the old

This ancient hoard of more than 6,000 third century bronze coins *(top)* was found by Ron Herbert *(center)* of Leicestershire, England. Ron was searching a field with his *GTAx 550* detector.

They were contained in an old pot, the top of which was buried 28 cm below the surface. *(Right)* The coins remain packed in the pot after it was recovered.

(Below) Closeup photo of one of the Roman coins which were reported to the local FLO for processing.

Ian Botley of Shropshire County, England *(right)* poses with a hoard of more than 2,800 fourth century bronze Roman coins he discovered with his *GTI 1500* detector. The Shropshire Field Liaison Officer (FLO) in his initial examination indicated that these coins *(above)* were from the 350–353 A.D. era of emperors Decentius and Magnentius.

bucket, he began pulling out masses of coins and knew that he had hit pay dirt. The emmar contained 1,327 coins, most of which were silver that dated from the 1950s back to 1860. Many of the coins were stamped with King Willhelm III, the late King of Holland. All told, the giant cache was valued at more than 30,000 €. This hoard also included four gold coins. His discovery was written up in the local press and the majority of his share of the find is on display in his showroom to this day.

Franco Berlingieri of Belgium had a similar experience finding hidden money for a family—a cache worth more than 50,000 €. He located some coins in a field near the home on his first hunt but he decided to come back again with his *TreasureHound* two-box coil to search deeper for a large target in its All Metal, Non Motion mode. He and his hunting partner first found the remnants of an old pot that had been shattered by a plough in the field. The coins

Two photos of another impressive hoard of coins found in an old iron pot by a detectorist in Europe with a Garrett *GTI 2500*.

from it had been scattered about the area, but Franco continued to dig deeper and scan. Beneath the first pot's remnants, he found an intact second cache pot that was filled with Hispania coins from the 1560s to 1570s. The pair recovered more than 20 kilograms of silver coins (nearly 400 total coins) and 30 gold coins.

Franco was also able to find another cache under different circumstances within an older home. In this case, a Belgian man had passed away and his estate was left to his cousins, his closest living relatives. They contacted Franco, offering him 10 percent of any money he was able to find. This particular deceased man had great means but had cached away his goodies, living more like a beggar. Franco was able to recover more than $500,000 in francs with his Garrett detector, much of which he discovered hidden in a special stash in the home's chimney, cemented away behind the chimney's inner walls.

Not everyone is so fortunate as to be offered the chance by a family to help search for a hoard where there is at least some information as to its whereabouts. Plenty of other grand cache pots have been unearthed in old farming fields where an ancient community once stood.

Tips on Cache Hunting

Many of the fields in Europe have rich history and have been worked in some cases for thousands of years. Here are a few basic tips to improve your chances of finding a cache:

- *All large targets should be dug.* Some of the larger cache targets may register on your detector as an overload signal. Many of these may prove to be old farm tools such as plough heads, but you just never know when one of those signals might be the tell-tale sign of true pay dirt!
- *A two-box searchcoil setup, or Depth Multiplier, is ideal for sweeping an area for large, deeply buried targets.* This special detector attachment can multiply by several times the depth-detecting capa-

bility of a compatible metal detector. More importantly, the Depth Multiplier ignores small objects, thus enabling it to overlook small pieces of shallow junk metal.

- *Keep an eye out for broken pottery, lead or other evidence that people had once lived or concentrated in a particular area.* Farmers often plough up bits of old pottery. These pieces can often be spotted on the surface as you sweep the fields. Areas where significant shards of pottery have surfaced are often very productive areas to find other metallic artefacts.
- *Be careful not to set your discrimination too high.* Some of the best ancient hammered coins are very thin and very small and could easily be missed using higher discrimination levels. These deeper, small targets often produce only a very faint signal and can often register on your target ID scale as tin or pull tabs.
- *Headphones are always recommended* to help block out external noise and to enable you to pick up the faintest signals of really deep treasures.
- *Keep an accurate record of the numbers and locations of coins and objects you discover.* Your journal may point out trends regarding certain areas that could benefit your future hunts.
- *Conduct proper research to determine the areas where old hoards were most likely to have been hidden.* Travel and research can be expensive, so it is good to concentrate on more than one area at a time to increase your odds of finding a healthy cache. You can never know too much about your target and the individual(s) who hid it. Talk to older residents in the areas where you plan to hunt. Some of them may share interesting stories—such as where their grandparents ran a country market—that could lead you to significant treasure recoveries.
- *Search all areas of older homesteads including inside the home.* The attic, basement, garage and barn areas are common spaces that occupants may have stashed coins, jewelry or other valuables.
- *Be prepared for the geographic conditions of the area when you begin your physical search for the treasure.* Ground mineralisation in certain regions may necessitate a different type of metal detector

Gilles Cavaillé of France was called upon by a relatives of a recently deceased man to find these eight 1-kilogram gold ingots he had hidden in his home.

than one you would use in a freshly ploughed farm patch. If heavy vegetation exists in your search area, use a weed cutter to allow your searchcoil to work close to the soil.

• *Again, larger searchcoils are better for locating a large treasure cache.* Silt, erosion, fill dirt and farming efforts may have concealed your target even deeper than you might have expected.

• *Finding a coin cache takes great patience and diligence.* Some detectorists spend years in search of a big haul, but those who reap the rewards are generally the ones who stay after it. True, chance discoveries are sometimes made by beginners, but you should never assume that a particular location should be passed over simply because it has been worked before.

• *Advertise your services and negotiate an agreement.* Before you start hunting, agree to terms with the property owner who might have a legitimate claim to your cache. The best, and most binding, agreements are always those that are put down on paper. Be sure to check the antiquity laws in the country where you are searching. In some countries, the law specifically states what percentage of a recovery is properly due to the land owner.

- *Deeply buried caches will require proper tools for recovery.* Most detectorists will experiment and settle on the digging implement that works best for them but a sturdy, steel shovel or spade with a foot peg for rugged soils will generally serve you well.
- *Think like the person who buried the hoard.* Simply put, practice burying your own cache on your property or on your farm. Where would you hide your most valuable possessions? What type of container would you bury them in? You probably would not want to bury your money or special items in broad daylight in an open area where your actions could be observed by others.

Many hoards are encountered by detectorists by pure chance. Such was the case of Russian hobbyist Denis Sokolov shortly after he decided to purchase a Garrett *ACE 250*. He made a few searches to familiarize himself with the metal detector before venturing to a farmer's field in an area southwest of Moscow some distance from the Kaluga Road. This ancient roadway had been in use since the Middle Ages as a major route between Kaluga and Moscow.

"My father decided to go with me to search with my detector in the field this day," he said. "He joked about finding a treasure chest full of gold coins." Denis searched the field for some time without luck before he and his father decided to head for home.

"On the way back across the field, my *ACE* gave a strong signal. The Target ID showed silver, so I began to dig," Denis said. "I picked up several clods of earth, bringing them to the coil. One rang. I broke it into two parts and one of the halves began ringing. From within the clod I suddenly saw a flash of silver. It was a coin!

"It was very unexpected. In my hands was a very small coin that was clean and bright. I just managed to see the emblem of

Moscow and St. George on horseback with a spear. On the other side of the coin were words written in old style letters.

"We were both very happy and it was an unusual feeling," he continued. "I knew that you could find coins in the fields but when my first discovery was a silver coin, it just seemed too sudden. Of course, at the time I made this discovery I did not at first realize that this coin dates back to the time of Ivan the Terrible in the 1500s. I was just happy to find it. We searched around a little bit more and found two more silver coins, so our catch on the first day was three silver coins of the same design.

"We went home and were able to identify our coins as being from the period of about 1500–1600 AD. The coins had lain in the ground for centuries as the history of Russia had unfolded. I could not sleep well that night. I had read on the forums that these coins were likely ancient treasure that had been disturbed by the farmer's plough. The fact that they were found so close together might mean that they were part of a bigger treasure. So, we decided to return again the next day."

Denis and his father returned to the same field and conducted a tight search in the area where they had found their first three

Denis Sokolov, a Russian detectorist who had recently purchased a Garrett *ACE 250,* found these 42 silver hammered coins from a hoard that had been scattered by a farmer's plough. They date to the reign of Ivan the Terrible in the 1500s. *Images courtesy of Denis Sokolov.*

Ivan coins. "Within an hour, we found 15 more of the exact same coin. The fact that this was a ploughed treasure was now obvious. All of the coins were found within an area of about 8 x 8 meters. The main concentration of the same coins were found in an area of about 3 x 3 meters. It should be noted that these silver coins are very small and are often hard to detect. Altogether, I made seven trips to this field. Once it became hard to get more signals, I would remove the top layer of the earth and then scan deeper. Applying this method, I managed to find another dozen coins, bringing my total to 42 silver coins from this Middle Ages hoard.

"I think it almost miraculous that I came to find this spot in the middle of such huge fields. I was lucky. Finding this not only increased my passion for metal detecting but also made me think about the fact that the mere acquisition of my detector enables me to find such treasures of our ancient history."

CHAPTER 7

BROOCHES, BUCKLES AND BUTTONS

"I've long been fond of vintage pieces," admits Russian metal detectorist Maxim Burmistrov of Moscow. As a child, he enjoyed the details of old-fashioned items such as a nice knife, a fountain pen or his grandmother's old coffee grinder. One of his most prized heirlooms is the *samovar* (a metal urn with a spigot used to boil water for tea) purchased by his great-grandfather many years ago with his hard-earned money.

As an adult who is raising two sons with his wife, Maxim now tries "to instill in them a love for the history of our family and of our country." His love of vintage items recently led him to purchase his first metal detector. His first hunts were made in the fields just beyond his home on the edge of Moscow. He found a number of old coins on his first day of searching.

Maxim found more coins, bullets and other items on his next searches. "I never could have imagined that the earth conceals such a variety of metal!" He turned to the Internet for tips on good places to hunt and read that old churches were an excellent place to search. He recalled an old church with a bell tower that had been severely damaged during World War II. The area where it once stood had since become a summer camp for the church.

These grounds were not designated as an historical monument and were not otherwise protected by the state. Maxim searched the grounds of the demolished church several times and began to turn up beautiful finds, including a rare 15th-century icon.

Maxim next searched a village area which contained the church where one of his sons had been baptized. "The village is old and the temple was abandoned only a few years ago," he said. "I walked for five hours at a considerable distance from the temple, on the edge of this village and found nothing. Suddenly, I caught a sharp signal and began to dig."

He dug a religious decorative piece containing beautiful patterns and colors. Maxim found a number of interesting crosses, pieces of icons and various religious items during the next few months of searching. "Some time later, I received a message from

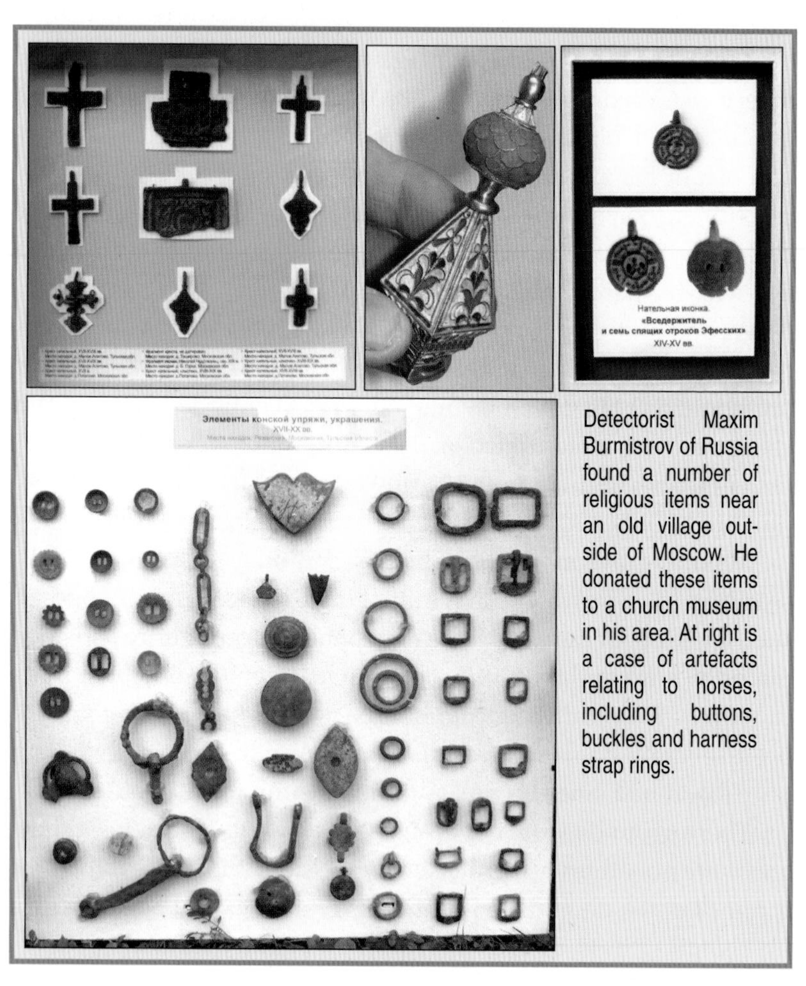

Detectorist Maxim Burmistrov of Russia found a number of religious items near an old village outside of Moscow. He donated these items to a church museum in his area. At right is a case of artefacts relating to horses, including buttons, buckles and harness strap rings.

the father at the church that he was opening a museum in the church Sunday School building," he said. "With joy, I took all of my previous religious item discoveries to donate to the museum. I was happy to part with my findings for two reasons. First, now they can serve young people and secondly the religious pieces came back where they should be."

The church museum included a section on horses and riding so Maxim also donated another display case dedicated to artefacts of that nature which he had found. Many of these finds were buckles, harness loops, buttons and other metallic remnants from saddles and harnesses used on horses hundreds of years before.

Maxim believes that by putting such artefacts on display he is giving them "a second life, perhaps much longer than their first life. Thanks to this hobby, I have found new friends, enjoyed new experiences, and have learned a lot about Russia's history. I never look at the search as a source of income. I like to save old things from centuries of oblivion and show them to other people. I have found a new passion in life which I hope will never leave me."

Some of the more common artefacts to be found on Roman sites in Europe besides coins are buttons, bronze brooches, buckles and fasteners of all types including those of small animal figurines. Metal buttons, used as practical fasteners and for decoration, date back to at least the Bronze Age (2100 to 700 BC). The majority of metallic buttons recovered by detectorists are of more recent origin—generally dating from the 18th to 20th centuries.

Soil acids and time cause most excavated buttons to lose their surfacing through deterioration, thus making the recovery of a decorative button in perfect condition a rare occurrence. Buttons recovered which are made of precious metal are more scarce, and may command a nice price from a collector.

During the Bronze Age, conical bronze buttons, about 2 cm in diameter, were used in contemporary dress. The surfaces of

(Above) These early silver and brass buttons, dating from the 1600s to 1800s, include fasteners worn by Dutch army soldiers and settlers. Such buttons from this period were often richly decorated with flowers and geometric shapes. Other buttons from this collection are show in detail in the photos below. *Courtesy of Leo Kooistra.*

such buttons are generally smooth with a face that raises almost to a point. The backs of these bronze buttons were generally very roughly cast and included a raised section through which a thread or leather thong could be passed to sew the button to a uniform.

Also found in Europe are decorative Celtic pieces made from wrought iron and bronze, with elegant Greek-influ-

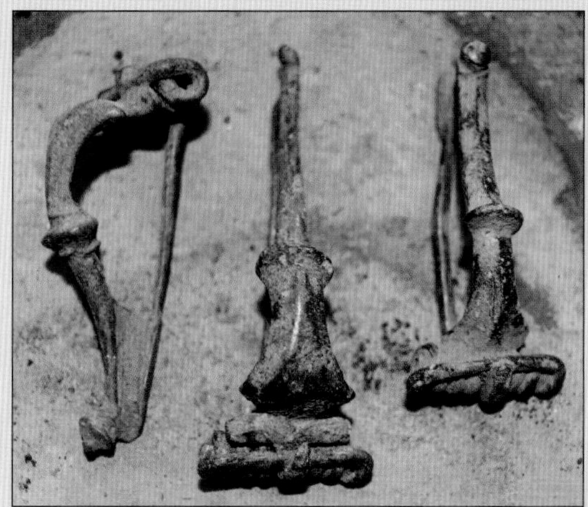

These three bronze Roman fibula brooches would have been used by middle class citizens around the second century AD to fasten their cloaks. Measuring about 6 cm, these fibulas were found with a detector in Belgium.

Several common bronze fibula and a Roman key, circa first–second century AD. The rounded fibula, a special clasp to hold a shawl or scarf around the neck, dates to about 500 AD.

ence designs, that date back to the Iron Age. Such pieces of Celtic influence are found mainly in France, Spain, Belgium, England and Ireland. Early Celts used a variation of the button known as a bronze "toggle," a fastener of varying shapes used to secure a cloak. Such Celtic toggles have been found in rectangular, square, triangular and butterfly-shaped forms. Early Saxon and Viking people even used bronze hook fasteners for their clothing.

This crossbow fibula was worn by a Roman centurion, a high-ranking officer in charge of at least 100 soldiers. It measures 9 cm in length by 5 cm in height and weighs 34.6 grams.

(Right) During the 16th century, a hooked belt fastener in the form of a serpent became popular in Europe. Such ornaments were often made of bronze or copper alloys and were used as part of a sword belt.

(Below) This early single-piece bronze fibula brooch is quite nice because of its arrow-like catchplate with intact pin. It is believed to be of Celtic or Etruscan origin from around 2000 BC to 1000 BC.

By the mid-13th century, a guild of button makers was established in Paris and decorative fasteners were more commonplace. Many were cast of pewter, brass or bronze but the elite acquired more pricey fasteners formed from silver or gold. “False” buttons were often sewn onto various parts of garments and even on hats. By the 18th century, copper had become one of the primary metals used in casting common buttons.

Bronze brooches, known to many as *fibulas,* date back to the pre-Roman days in Europe and can be found in all areas of Europe. Fibula is a Latin term to describe a fastener. Many such brooches were lost due to a damaged or broken pin, making the discovery of a fibula brooch with an intact pin particularly special. It was during the Bronze Age that brooches were first used as a functional dress ornament in Northern Europe. As the Roman Empire spread into northwestern Europe around the period of 50 BC, Roman fibula brooches became common in areas such as Austria, Belgium, France, Spain, Germany, the Netherlands and later England, as the island came under Roman control. Roman fibulas, generally between three to eight cm in length, were mainly made of bronze although detectorists can find simple iron brooches and the occasional rare silver or gold fibula. The first known fibulas made their way across the Channel into Britain by the eighth century BC.

Fibula brooches of early Europe were made of bronze, iron, bone, or even silver or gold in the more elite classes. These early fasteners were cast as a solid piece with a hinged pin that was used to fasten a heavy woolen cloak. Fibulas became more sophisticated as the centuries passed, and bilateral springs were added to such fasteners versus the straight hinged pin. Women used brooches to fasten tunics as well as cloaks.

More ornate brooches known as *plate brooches* achieved their peak of popularity in the second century AD, although they continued in use throughout the Roman period. Plate brooches are often larger disks that contained more ornate designs that included beads, glass inlay, painted colors and special designs. Others were crafted into the form of turtles, snakes, rabbits and other figurines.

(Right) This early broach included cut glass stone ornaments.

(Below) Ancient Roman fibula brooches found in England.

(Left) Front and back side of a unique Roman fibula from the second century AD.

Yet another fibula style is the *crossbow brooch* which was in use by the third century BC. This fibula resembles a highly arched bow with a long catch plate recessed for securing the pin. These crossbow fibula brooches were increasingly heavy by the fourth century as they were made larger to hold heavier clothing; artisans therefore began making hollowed crossbow fibulas by the end of the 4th century to cut down on weight without sacrificing their ability to secure delicate fabrics. Another special find for the detectorist is a *rounded fibula or ring brooch,* which consists of a ring of metal fitted with a swivel pin. The pin is trapped in place by a recess in the metal at the point on the fibula where the pin hinges. Ring brooches date back to at least the 12th century but became very popular during the 13th century. While commonly made of bronze or pewter, some rounded fibulas have been excavated that were made of silver with detailed inscriptions and patterns. These particular fasteners were commonly worn to fasten undergarments near the person's throat.

Roman- and Saxon-era straight pins were also used to help pin garments and as hair pins. Roman ladies wore their longer hair rolled into some form of a twist and secured with such hair pins. Those most commonly found are made of bronze, some containing ornate designs on the pin's bulbous head. Excavations also turn up more recent straight pins made of copper alloys during the medieval period.

European detectorists often discover buckles of all shapes and sizes in the field. It is often difficult to determine the age of the more simplistic, plain buckles that are unearthed but such fasteners were used from the Roman period to the medieval period and into more modern times. Buckles of brass, bronze, and silver may be found in all types of designs and sizes. They were used equally for securing pants and blouses and as ornamental pieces on shoes, uniforms and even hats.

(Above) These handcrafted silver buckles of all size—for shoes, shirts, blouses, and pants—were found with Garrett detectors in Europe.

(Right) This photo shows the typical condition of an early buckle that has just been unearthed in a recently ploughed field in Belgium.

(Above, left) This silver buckle and the shoe buckle *(above, right)* each date back to about 1650 to 1750 AD.

This special "mille fiori" Roman fibula *(above)* dates back to the first century AD. Although it is quite worn from many years in the ground, you can still see the fine details of one of the tiny flowers (circled) which once decorated this piece. *Courtesy of Franco Berlingieri.*

(Above) Various early fibula found by a European metal detectorist. He keeps these in a display box but considers them less valuable because they do not have their original clasps.

(Above) Detectorists are thrilled to find complete sets of multi-part buckles. *(Left)* This French buckle, known as a *plaque-boucle circulaire*, is a Celtic bronze piece used around the seventh or eighth century AD during the Mérovingienne period. It was found incomplete. (See below for a complete three-piece set of such a buckle.)

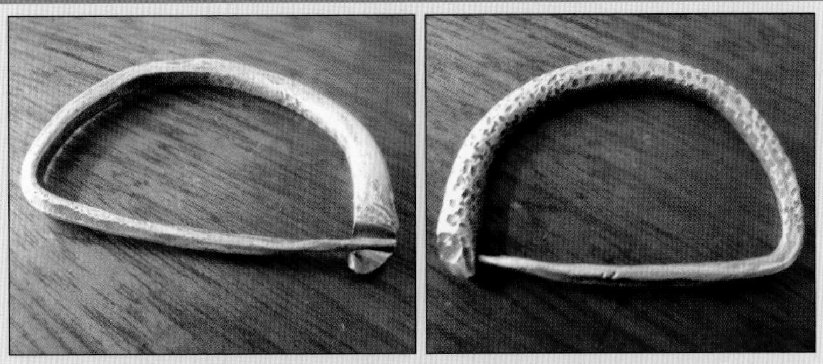

(Above) This silver crossbow fibula broach, circa 1000 BC, was found in France. A silver fibula is quite rare and is a treasured find for any European detectorist.

(Above) More early buckles and other artefacts recovered by French detectorists. Clockwise from upper left: a Medieval period méreau, a church token given to the poor for food or clothing; a copper medallion from the 1800s era; a Roman bronze razor; a large bronze belt buckle from the Mérovingienne period; a Medieval bronze chain for military decoration; and two bronze Medieval period buckles, smaller for clothing fasteners. *Photos courtesy of Gilles Cavaillé, Loisirs Detection.*

Gallery of Finds from a French Detectorist

(Above, left) Louis XVIII bronze clothing button. *(Above, center)* 18th century tunic button. *(Above, right)* This bronze piece dates to about the 13th or 14th century AD.

All photos on this page courtesy of Regis Najac.

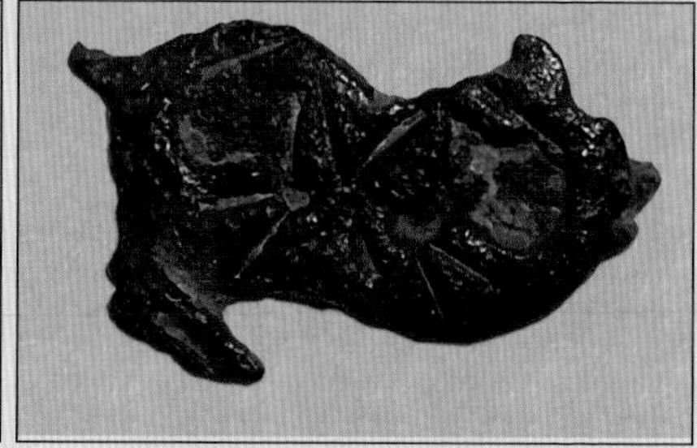

(Above, left) This bronze Gauloise button, circa 150 to 100 BC and 2.7 cm in diameter, was the first loop of an ancient belt loop. *(Above, right)* A bronze Roman Zoomorphe fibula approximately 2 cm in width.

(Below) Regis holds out a beautiful 19th century bronze and marble bracelet he recovered.

CHAPTER 8

RINGS, JEWELRY AND FIGURINES

Nigel Ingram's favorite finds with a metal detector in England have been given away. Actually, they have been *returned* to their owners.

Nigel and his father have been called upon a number of times to help find family heirlooms, valuable jewelry and other trinkets that held deep sentimental values to their owners. In most cases, they are more than happy to be rewarded with a cup of tea and a hug from the overjoyed person who has been reunited with their jewelry. When he was only 14 years old, he received a reward that was worth more than the Garrett *Groundhog* metal detector he was using at the time.

He was already a seasoned veteran of detecting. Nigel started going out hunting with his father Derek Ingram at age 8 in the outskirts of Birmingham, England. Derek, a former Royal Air Force pilot, had become a metal detector dealer who spent a good deal of time in the field testing his machines. "One of dad's first detectors had a wooden search coil," Nigel recalled. "It was built by Mike Beach, who later founded M. L. Beach Detectors in the UK."

Derek often took his sons Nigel and Marcus along on his hunts. "My father had found a ring on a farm that the farmer's wife had lost," Nigel recalled of one particular search. "Pop was having a cup of tea with the lady when she related that one of their customers had lost a ring in their strawberry patch." The Ingrams went back and were able to find this lost ring with their metal detectors.

"Fruit fields are a great source for finding rings and coins lost by people who are rummaging about picking fruit," Nigel offers. The farm owner was so amazed that the trio had been able to find both lost rings that another story happened to come out in conversation. "One of the farmer's wealthy relatives had lost a valuable family crest ring 17 years ago while he was helping to toss bales of hay into the field from the back of a truck."

The Ingrams followed the farmer and backtracked along the route of where the ring had been lost nearly two decades before. In just over 90 minutes, Nigel located the large gold ring with his Garrett *Groundhog* detector. "It was 22k gold worth £12,500 in the 1970s and was large enough to fit right over the wedding ring on the farmer's hand," he recalled. "His relative must have had fingers as thick as frankfurter sausages." The farmer returned the family crest ring to his relative in Belgium, who sent a reward to young Nigel.

"I wasn't expecting anything more than a smile and a 'thank-you' from the owner," he stated. "The reward was more than what

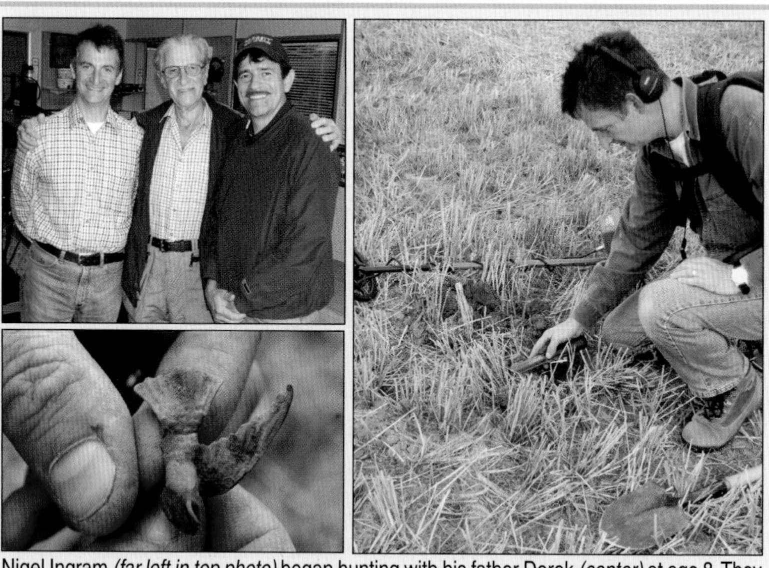

Nigel Ingram *(far left in top photo)* began hunting with his father Derek *(center)* at age 8. They are seen with Garrett's Henry Tellez at Regton, their Birmingham business. Nigel uses his *Pro-Pointer* to find an interesting bird pin lost in this field many years ago.

a new detector would cost me, so I was thrilled to bits." In addition to recovering lost rings, Nigel has found watches, necklaces and various other gold and silver European jewelry over the years. "My motto is 'slow and low' for making good finds," he says. "To be the most efficient, you want to keep your searchcoil low to the ground and be careful not to swing too fast."

Decorative pieces—rings, necklaces and other ornamental bodywear—often indicated a person's wealth. The metal content of such items can be indicative of the period, as can be the particular designs in some items. This chapter is designed to reflect the wide range of jewelry and collectibles that has been unearthed

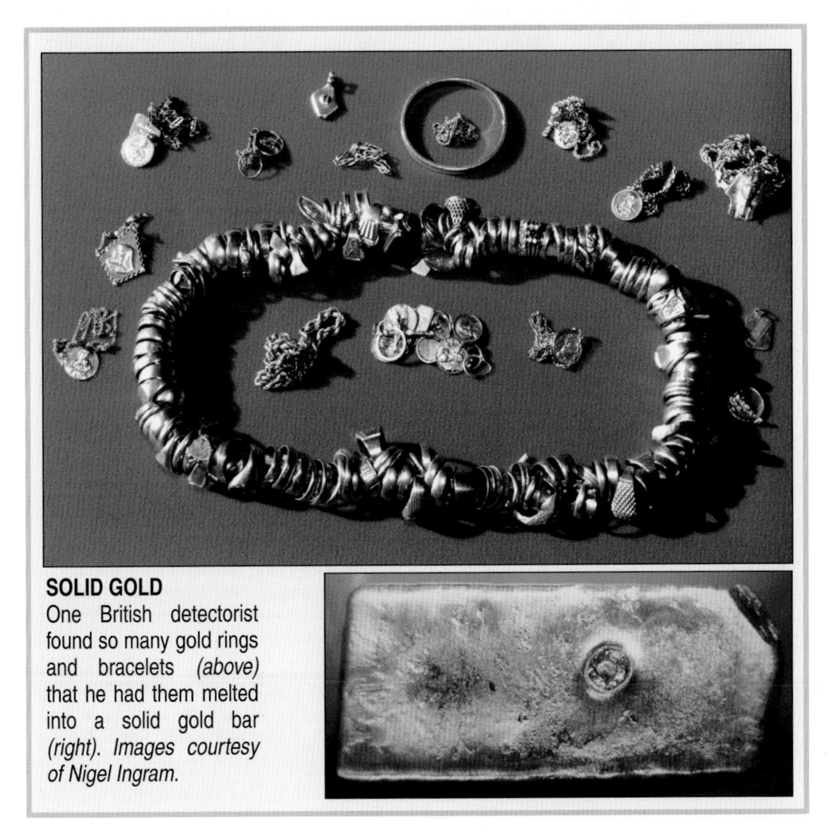

SOLID GOLD
One British detectorist found so many gold rings and bracelets *(above)* that he had them melted into a solid gold bar *(right). Images courtesy of Nigel Ingram.*

These 2,000-year-old bronze rings each bear the owner's personal insignia, which could be used to officiate a document. Two of the four rings shown below are seen in more detail to left and right.

(Left) Bronze Roman "S" symbol ring.

(Below) Some of your more modern jewelry recoveries can be just as valuable as the old pieces. This stunning 18k gold watch, weighing 61.3 grams and containing 110 diamonds, was found by a Garrett detectorist on a European beach.

(Above) Found near the oldest Roman settlement in Belgium, this bronze ring would have been worn by a person of royalty or with stature in the church.

with a metal detector. Hobbyists have found rings and other valuables that were lost since they first began being made.

Modern watches, necklaces and rings range from the ordinary to the extraordinary. Carry your detector with you on vacation to the parks, the beaches, the mountains or wherever you may journey. You can bet that there are plenty of good items to be found.

The metal content and the design style of ancient treasures can be indicative of certain periods of history. Rings from the Roman era were most often made of iron or bronze. The detectorist who digs a gold or silver Roman ring should consider himself lucky, as only the upper class of society could afford these. Various stone settings or glass ornaments were set on some such rings in lieu of more expensive precious stones.

Silver was used more and more by the seventh century as the European gold supply dwindled. Bracelets, hairpins, necklaces and other ornaments of gold and silver can be found that date back many centuries. By the Middle Ages, medieval rings in Northwest Europe were often very plain and were made from bronze, brass or copper. The fancier ones might contain an inscription or a personal seal on the ring's top face. Such *signet rings* were used by the owner to stamp his personal seal into important documents or correspondence. Those in the upper class had more highly decorated rings of gold and silver which might contain precious stones and intricate patterns.

The more basic, smooth ring of this era has been referred to as the *hoop ring*. Other rings from this period might have been simple, twisted brass pieces. *Love rings or posy rings* became popular in Europe toward the end of the 16th century as signs of affection. *Religious rings* were also popular during the medieval period and some wearers no doubt believed this piece of jewelry might serve as a protection piece against evil spirits. Rings were used for a variety of other purposes in ancient Europe. They not only were as decorative signs of the owner's wealth but were given for medicinal, political, ceremonial and mourning purposes. In some cultures, mourning rings were given to friends and family of a de-

The gold and saphhire Middleham Jewel, found by a Garrett detectorist, is one of the most valuable pieces of European jewelry ever recovered with a metal detector. The front of the pendant *(left)* depicts the Trinity scene while the back *(right)* shows the Nativity.

ceased loved one. These rings were often inscribed in the memory of the departed person or contained special symbols, verses or other inscriptions.

Some lucky detectorists have made truly remarkable jewelry recoveries in Europe. Ted Seaton of England, for example, found a precious religious pendant believed to have belonged to King Richard III. This 15th-century treasure, known as the "Middleham Jewel," is perhaps the most rare and valuable medieval jewel ever found in Europe. Seaton found the pendant using a Garrett metal detector at a depth of 14 inches along a well-worn footpath near Middleham Castle in North Yorkshire, England. This castle was the boyhood home of King Richard III. Seaton had been preparing to quit searching on a rainy day when he detected the faint signal. He dug the item and stuck it in his pocket, assuming at first that it was a lady's compact. He was quite shocked when he began cleaning the 69-gram medallion to see its detailed golden surfaces and a 10-carat blue sapphire.

Leo Kooistra of Holland *(above)* has found literally thousands of silver rings with his detectors, such as these fine examples *(below)*.

(Upper left) Silver crosses found in the Netherlands by Leo.

(Left) Medieval crosses, circa 1750 AD, found in Belgium.

The 6.6 cm tall by 5.9 cm wide pendant contains the large blue sapphire on the front and intricate religious engravings on both sides. The Middleham Jewel, judged not to be a Treasure Trove, was auctioned in 1989 at Sotherby's of London for £1.3 million. The British public was disappointed to find that the unique jewel had been allowed to leave their country. Some ten years later, however, the Yorkshire Museum in York raised £2.5 million to buy back the exquisite Gothic pendant.

Albino Bartolini, who lives on a high hill overlooking the breathtaking Mediterranean coastline of Citavecchia north of Rome, has been a treasure hunter for four decades. He has found rings and jewelry that many of us could only dream about. He is quick to point out the importance of using a quality detector that is able to quickly recover from the masking effects of an iron object located in close proximity to a treasure target.

Bartolini believes that early Europeans generally hid their items of value within 200 feet of their home. He looks for stone or terra-cotta remnants and then sweeps out from the area where an early home once stood. "Sweep your searchcoil at an angle to the place where the home's walls once stood," he advises. "In Italy, it is possible to find good treasures almost anywhere. You have 8,000 years of history in my country, so there are good things to find almost anywhere."

Bartolini advises that detectorists should be aware of the local laws where they hunt and should conduct proper research to help locate good hunting areas. "For those just getting started," he says, "I recommend that they spend some time hunting with someone who is very good with their detector."

He has found many gold coins during his years of metal detecting, including a gold 50-liro Mussolini coin that was in near mint condition. His favorite jewelry recovery is a gold papal ring dating from about 1300 AD, which was found in a field in northern Italy. "It was buried deep in the ground, about 9 inches," he said. This valuable ring had been carefully cached away by someone who surrounded it with stones in the fashion of a protective box.

Even children have left treasures that people find today. A Russian detectorist, Andrey Shipko, found such a cache near the remains of an old farmstead that had been destroyed in the 1920s. He and his father had accompanied their friend Eugene to the site where he had found coins, buttons and silver dishes in the past.

"We agreed to give him one-fifth of whatever we found with our metal detectors," said Andrey. En route to the site, he and his father Anatoliy found several copper coins and a silver 10-kopeck coin from 1927. The farmstead they reached had only the foundation remaining where five wooden buildings had been.

By searching around the foundations and near an old overgrown pond, the men recovered various aluminum, brass and copper utensils, assorted agricultural tools, horseshoes and a 1868 three-kopeck coin. "When it began to rain, my father Anatoliy went with Eugene to take cover under a large oak tree," Andrey said. "He was scanning his GTI 2500 around the base of the tree when he got a large signal. He called me to take a look. The signal was very strong, although it bounced between aluminum and iron." Andrey dug up a rusted iron box about 22 cm in diameter.

"There were 15 kopeck coins from the years 1926 to 1929. There were also buttons and a copper ring with the year 1918 stamped on it. There were also pieces of colored glass and the remainder of an hour chain. It was a very uncommon finding. We agreed that it appeared to be the treasure of children and that they had buried it around 1929. It looked like the kids had saved coins that they were given each year along with other keepsake items. The child obviously buried his treasures under this large oak some time before his family's farm burned down. The earth holds many secrets but not all of them are happy."

These 1920s kopeck coins (left) and a copper ring from 1918 (right) were among the items found in the child's cache by the Shipkos.

Both of these bronze Roman rings date back approximately 2,000 years. The ring seen at left contains a small hematite stone on top while the ring to right contains a simple green stone in its mount.

This Garrett detectorist's display case contains gold, silver and bronze jewelry found during 25 years of searching in Europe. Some of the older, intricately-detailed Roman and Celtic rings date back more than 2,000 years.

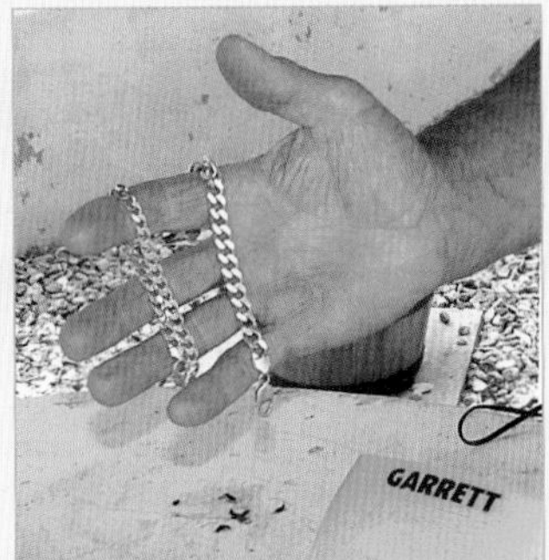

Jack Dey *(left)* of England took his new *ACE 250* to the beach to search along the tide line. During his first outing, he found coins, two Sterling silver bracelets *(see below)*, and a gold ring with 12 diamond chips and a central amethyst.

(Below) An ornate pill box found in Belgium.

(Right) This gold signet ring was discovered by French detectorist Regis Najac.

(Above) This collection of rings covers a wide range of French history from the Celtic age to Roman times to the French Merovingien period (approximately 600 to 800 AD). Courtesy of Bruno Lallin, Lutéce Détection.

Regis Najac has recovered both ancient and modern sporting medallions with his detector. These ancient Roman medallions *(at left and at right)* date from about 100 BC to 100 AD. Note that the more modern sports medallions, created in the 1800s, very closely mimic the original Roman sports medallions.

Modern jewelry can be found with metal detectors almost anywhere people have been swimming or playing outdoors. These are just a few of the hundreds of gold rings that Leo Kooistra has found in Europe.

(Left) A Roman-era bronze ring, circa 100 to 200 AD, shaped like a coiled serpent. *(Right)* This bronze pieces of Roman erotic art were used as fertility charms. For example, a family experiencing fertility problems might be given such a charm as a sign of good luck or hope.

Albino Bartolini *(above)* found this gold papal ring *(left)*, circa 1300 AD, buried deep in a field in northern Italy.

FINDING HISTORY: Adventures of a European Detectorist

Some of the best recoveries made by European metal detectorists are far off the beaten path. Such is the case of an adventurous detectorist who goes by the nickname of Manuds. He was happy to share the following story of his greatest discovery to date—a small Roman bronze sculpture that dates back to the first century AD which he found in the Pyrenees Mountains in France.

"My friend 'P' heard a story over dinner one night from a friend of his buddy's parents. The man said that his grandpa used to go to the Pyrenees Mountains in France to look for treasure near a little mountain called Gardillet.

"Of course, his grandpa did not have a metal detector in 1940 to search this area. He used a pendulum, which is very random, so I don't know if he ever found any treasure or not. My friend 'P' knew that this area of the Pyrenees had been occupied since the antiquity period by Romans and Celtics to dig iron from the mines. My friend kept this story and began searching with his Garrett *GTI 2500* detector.

"He found some bronze Roman coins but nothing extraordinary. He continued to hunt this mountain and on his sixth hunt, he found a silver Roman coin near the top of a hill. Using his *GTI 2500,* he suddenly began finding Roman coins everywhere!

"In a year's time, he found 300 Roman and Celtic silver coins. The grandpa was right! The coins were from the first century before Christ to the fourth century AD, meaning that the place was used for five centuries.

"In May 2009, 'P' decided to take some of his good hunting friends to this special hill to celebrate his birthday. We enjoyed three days of hunting with some good wine and some good Pyrenees food to survive.

"It was quite an adventure just

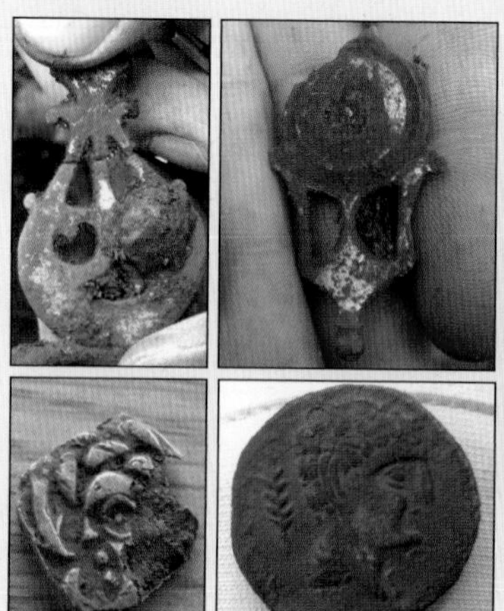

This treasure gallery represents some of Manuds' favorite metal detector recoveries. These medieval silver coins (above) date from 1100 to 1300 AD.

The last leg of these friends' journey to their special hunting area is a hike through the forest. Metal detectors and camping gear must be carried the last half hour.

getting to Gardillet. We traveled an hour on the main highway and then another hour on a smaller back road. The last hour of driving had to be with 4x4 vehicles on a little dirt mountain road.

"Finally, we had to carry our tents and backpacks for the last hour up a steep trail into the mountains. The top of the hill is 3,600 feet high in a beautiful forest on one side and a lovely valley on the other side.

"With eight friends, I hunted all over the hilltop and we found many coins, fibula and other Roman stuff. I found a silver medieval coin that is about 1,000 years old, meaning that medieval people once used this area to protect their frontiers. My best find was on our last day at this camp.

"I was hunting only about 100 feet from our camp fire, a place

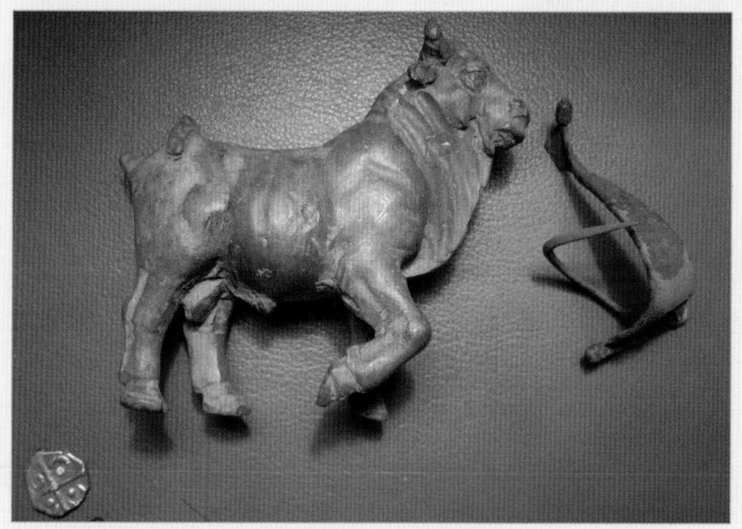

Manuds' favorite find has been this ancient 220-gram Roman bronze bull. The bull was once worshipped by Romans as an idol.

where everyone had hunted ten times already. Suddenly, *wiiing-wiiing!* a big noise came from my detector.

"I dug 10 cm and nothing. At 20 cm, I was just cutting through more of the tree roots with my knife. At 30 cm, I dug through some rock and then some clay at 40 cm. I used my *Pro-Pointer* to check the hole but there was no noise. What the hell? Where was the target?

"I looked at the big ball of clay that I dug from this hole and suddenly I saw this big, heavy, beautiful bull! What a masterpiece! I called 'P' to show him my new discovery. He came over nonchalantly, looked at the bull and then kissed my forehead. He was very happy and thanked me for this marvelous pleasure. All of my friends wanted to kill me with jealousy.

"We decided that this place must have been a Roman fanum, a temple where people used to go and give offerings to their gods. Roman people liked to give gifts to the gods, such as coins and fibulae. We also found some silver paper and gold paper, which was a typical offering, in the ground near the bull."

The Roman bull is about 8 cm (3.15") tall by 7 cm (2.75") long and it weighs 220 grams (7.76 ounces). Manuds and his friends continue to visit their favorite hunting ground when they are able. They continue to find fibulas as well as bronze and silver coins. "It's always a good trip to go there," he reflects, "but I'll always remember my best hunt."

(Below) Manuds also found these Merovingian belt parts, which date between 500 to 800 AD, in the Pyrenees Mountains.

CHAPTER 9

MILITARY ARTEFACTS

Danny Reijnders from Belgium has enjoyed searching for military artefacts for more than 15 years now. "It's almost like reliving the action when you come upon a significant battle area," he recounted. One of his favorite detecting areas was an area where Germans and Americans happened upon each other in the forest and got into a heated conflict.

"This battle is not published in books. It was really a lucky find coming across this spot," Danny admits. "We found American foxholes in the forest and began detecting cartridges and other artefacts in and around them. We looked at which way the foxholes were dug and scouted out in the logical direction toward where their enemies must have been. Sure enough, we soon found foxholes some distance away in the forest that were facing back toward the Americans. In this area and in between, we found German bullets and other artefacts. It was quite exciting."

Danny utilizes just enough discrimination to get rid of very small shrapnel and tiny iron items. "Some iron and steel items are the best artefacts, such as bayonets and rifles," he says. "You must keep some discrimination." Danny researches the areas carefully where he hopes to hunt and locates a landowner willing to give him and his friends the necessary permission.

He is now on his third Garrett metal detector, a *GTI 1500*. "I have found quite a few World War II items, including bullets, hand grenades, German bayonets, a German name plate, a com-

(Left) Artifact hunter Danny Reijnders, seen with his *GTI 1500* and the Mediterranean behind him, shows several coins and a Roman fibula he found in Italy in 2009. *(Below)* German cartridges, bayonet, grenades (all inert), ammunition magazine, case from a 75 mm howitzer gun ammo, signal cartridge, two buckles (one aluminum, one steel), and a telephone.

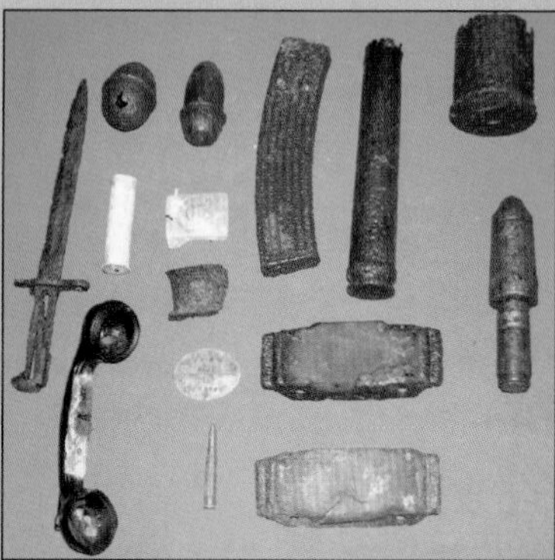

(Above) Some of Danny's English and American artifacts. Shown: pick-axe, axe, holder from Garand bayonet, three empty Garand clips, part of a MILLS grenade, gas mask, bazooka grenade fins, 2" mortar grenade fins, rifle parts, mess tin, equipment buckles, full Garand clip, safety pieces from English mortar round and airplane bombs, two U.S. shovels, mortar grenade and howitzer parts, .50-caliber cases, and 20 mm cases.

(Right) A Garand rifle, two German Mausers, Belgian rifle and two old revolvers (Belgian) all found by Danny.

(Below) Some of Danny's German artefacts. Shown: 105-mm howitzer case, helmet and gas mask, S (spring) mine pot and concrete mine (inert), tank-track, ammunition package cover, gas mask, MG ammo magazine, ammo package covers, hand grenade (inert); high pressure air container for life jacket from German pilot, shovel, part of a rifle, MG ammo on links and a MG ammo case.

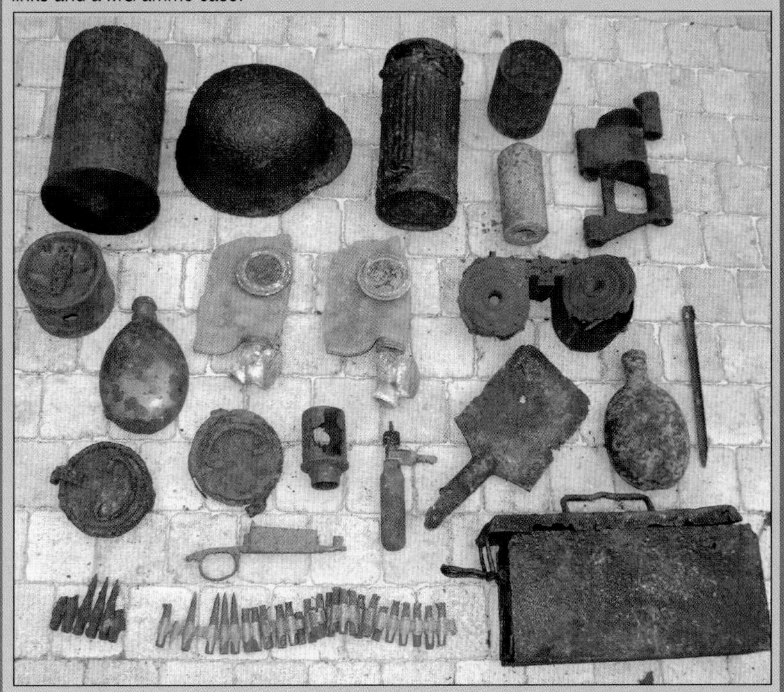

plete German gas mask, helmets from multiple countries, a rifle and two Mauser rifles.

"My advice," he says, "is to follow the laws about military artefacts. Many so-called war relics are now forbidden to collect. Know what the rules are before you go out. Aside from research, my best finds have come from being persistent and having a little bit of good luck."

Many detectorists are fascinated with finding military artefacts from wars as recent as World War II back to conflicts thousands of years ago. Each country in Europe has its own antiquities laws concerning items of historic interest. Be sure to consult Chapter 15 for European treasure laws—and seek specific current information in the country and area in which you hope to search—before going metal detecting. Violations of these laws regarding military and historic artefacts can be quite severe.

Weapons and bullets are key items that can be found at former battle sites, abandoned military posts, campgrounds or training grounds. There are many books and Internet sites that display sample items from various wars, regions and periods of history. This chapter will offer a general overview on artefact hunting, and it is illustrated with representative samples of military items recovered from different time periods and regions of Europe.

Musket balls, rifle balls and other round lead bullets were used from about 1500 through the 1860s. By the 1840s, conical-shaped lead bullets were in use for firing from percussion-cap weapons. Guns with rifled barrels have fired pointed lead bullets since about 1780 and the sophistication of bullets has progressed since that time. By the 1870s, centerfire cartridges which housed the bullet, gunpowder and the percussion cap were in use. The ammunition and the calibers range widely by the 1900s and many different military ammunition artefacts from World War I and World War II can be found across Europe.

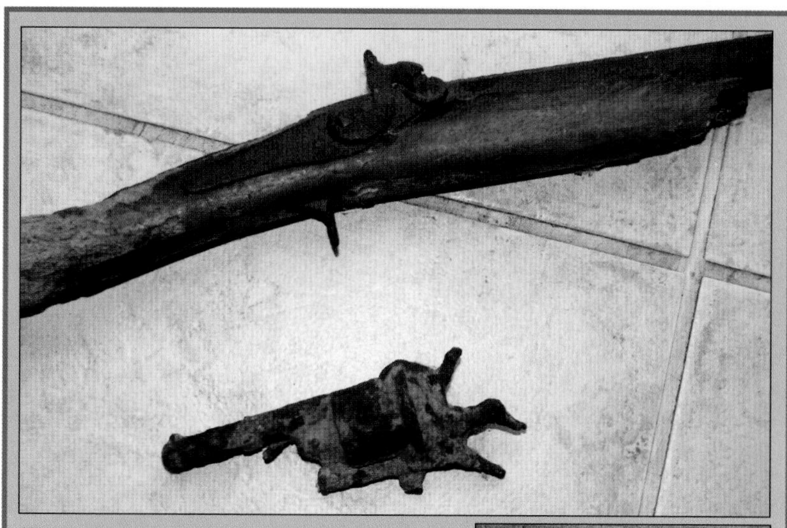

(Above) Deteriorated remains of early flintlock rifle and cylinder pistol found by European detectorists.

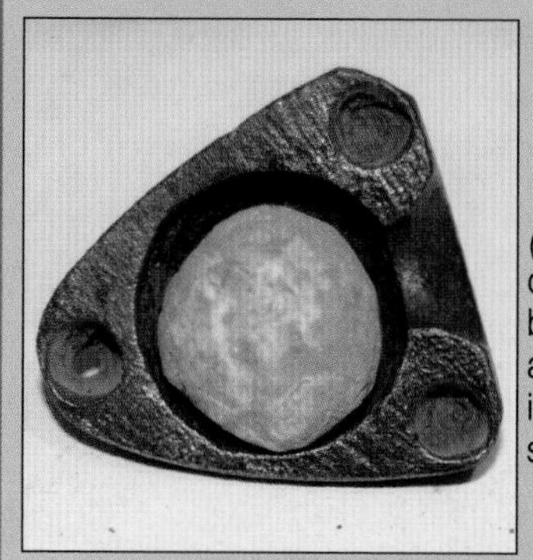

(Left) One half of a musket ball mold (with a musket ball inside to show scale).

(Above) These ancient bronze weapon tips were used by the Celts in Europe around 1500 BC.

(Left) This votive ax head, shown at actual size, was used by the Celts for trade or for offering around the period of 700 to 800 BC.

(Above) World War II artefacts recovered by Charles Garrett during some of his European metal detecting trips. The items seen include a German Army helmet *(left)* a British helmet, bayonet, knife sheath, a magazine of bullets, a first aid kit and a smoke bomb.

Prior to modern bullets or even those of lead, the earliest tools of warfare were swords, daggers, knives, axes and similar weaponry. In addition to ammunition and weapons, European detectorists search avidly for uniform pieces and other collectible military items. Uniform buttons, commendation medals, brass buckles, bayonets, ammunition belts and helmets are just a few of the many military-oriented items that are found each year with metal detectors. Again, it is imperative that you understand the laws of your country before engaging in this type of hunting.

One recommendation is to join a hunting club in the area you are hoping to search. Some clubs may partner with archaeological projects to lend their metal detecting skills. These objects are, of course, conserved by the archaeologists as historians, but the thrill of hunting an important battlefield is pay-off enough for some detectorists. Photos of your artefacts are a great way to document treasures that you are not allowed to keep.

Derek Ingram still recalls his younger days in Birmingham, England, during World War II. "The Germans bombed London

(Above) These cannonballs were retrieved by Derek and Nigel Ingram from an armed British yacht that was wrecked on the North Wales coast in the late 1600s. From left to right the weights of these cannon shot are 1.26 kg, 2.65 kg and 7.48 kg. The larger cannonballs were actually used as ballast in such early sailing vessels.

(Above, left) Derek Ingram used a Garrett *Groundhog* metal detector to retrieve these British 20 mm cartridges and Very pistol cartridge bases off North Wales in the 1980s. *(Above, right)* Fired from British Spitfire fighters, the 20 mm shells were date stamped for the military.

(Left) Nigel Ingram has found many World War II-era British .303 bullets with his metal detectors. The spent projectiles were both sharp-edged and blunt-tipped varieties.

(Right) Various early projectiles found by author Steve Moore while metal detecting in Europe. At top left is a standard French musket ball found near a skirmish area in Belgium. In upper right is an early .22 caliber bullet used in English muzzle loading rifles during the 1700s and 1800s. On the bottom row are two English pistol balls of about .40-caliber, one crushed by impact.

heavily but Birmingham, being the second largest city in England, was also raided quite a bit," he relates. One of the greatest dangers was the flying shrapnel from bomb explosions.

British metal detectorists such as Derek have found their fair share of such air raid artefacts in recent decades. He and his sons Nigel and Marcus have collected numerous .303 shells, the standard issue of British riflemen, from all over the United Kingdom. They have also dug plenty of piece of copper shrapnel from the rifling bands of large-caliber shells fired at attacking aircraft, as well as countless brass 20 mm cartridges fired by British Spitfire fighters at the Germans.

One of Nigel's favorite artefacts is the lid from a 1914 brass commemorative box engraved with an image of Princess Mary that contained clothes, candies, cigarettes and other items for World War I soldiers. The so-called "Christmas Truce" was a brief yet surprising truce between the British and Germans called at the Western Front. Troops emerged from their trenches to exchange gifts and some even participated in a multi-national football game. This unique truce was short-lived as the war returned in full fury the next morning but it is one of the more interesting pieces of trivia to come out of that world war.

Nigel's box lid from the 1914 Christmas Truce is certainly an interesting military artefact. Any such artefact—whether from the World War I era, medieval times or even Roman—tells a story of Europe's rich history.

(Above) This wide-ranging collection of military artefacts was found over many years by Nigel Ingram. In the upper left are a 20 mm shell casing, musket balls, a Very pistol cartridge cap, and a .303 bullet. Also seen at top are a mortar shell and the tail fins off a smoke shell. In the center is a 1915 World War I Hales Mark 2 rifle grenade. In the lower left corner are cast lead .45- and .67-caliber bullets and a rifle rest found at a World War I rifle range in England. The lower right contains various copper and brass World War II shrapnel. Also seen above and in this detail photo *(below)* is a Christmas 1914 brass lid from a gift box of personal items distributed to British, Russian, French and Belgian soldiers during the "Christmas Truce" of World War I which is described in this chapter.

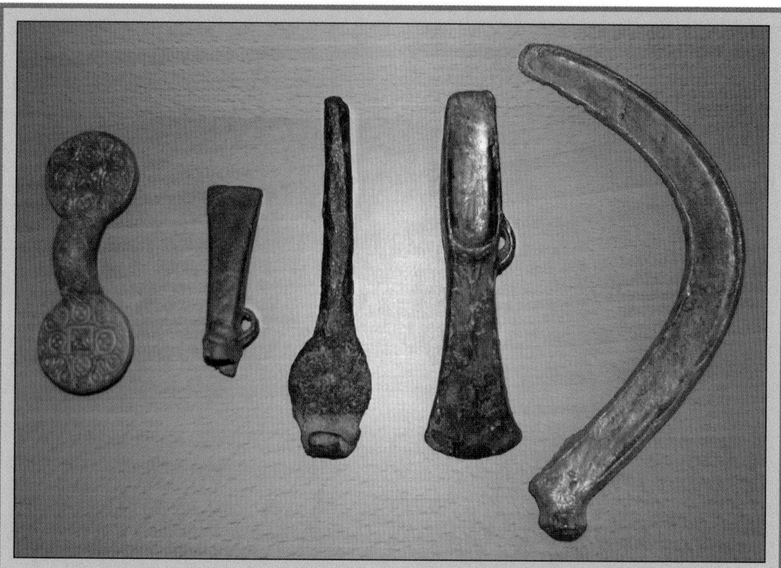

(Above) Ancient weapons, tools and a fastener found in France. Left to right: an 11.5 cm bronze fibula from around 600 AD, used by the Celtics during the Mérovingienne period to secure heavy clothing; 8 cm small bronze ax from the Mérovingienne period; 15 cm bronze tool; 15 cm bronze Celtic ax head from about 1500 BC; and a bronze crop cutting tool believed to have originated from the Celts.

(Left) This 13th century medieval crossbow arrowhead was found by Regis Najac of France. *(Below)* Regis also found this 13th century bronze and enamel *vervelle* (a small metal staple used in Medieval armor to attach a flexible curtain to a helmet) that depicts Richard the Lion-Hearted.

(Below) This tiny medieval artefact, likely either a votive axe or a pipe tamper, was found at a 2009 UK rally.

These oval rings of bronze and copper *(above and right)* were found by Regis Najac during his years of metal detecting. The photos were shown to treasure recovery expert Robert Marx, who quickly helped identify them.

"The smaller rings were part of the armor vests worn during battle and the larger ones were part of the harnesses for horses," wrote Marx. "I'm sure of this as I found the exact same goodies in Mallorca and several other sites in Spain, at Carthage, Tunisia, and at Tyre, Lebanon. At first, I had a hard time getting them identified."

The most impressive set of these rings that Regis found together (below) includes four bronze loops and one made of gold!

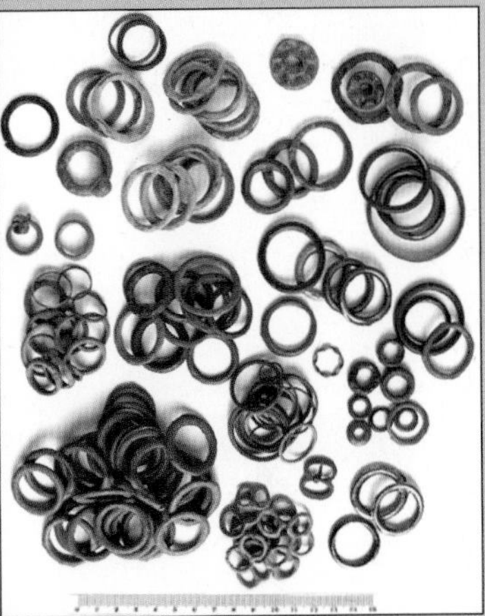

EUROPEAN RELICS FROM WORLD WAR I RECOVERED WITH METAL DETECTORS

(Right) These buttons are from the drill plate of a tunic, part of the uniform of a Bavarian light horseman. The top three buttons on the left, 23 mm in size, fastened the right sleeve of the soldier's tunic. Note that the lion faces in, with his tail on the left. The top three buttons on the right face the opposite direction. The two bottom buttons, 18 mm in size, also face opposite directions. The small photo below shows how these rows of lion buttons secured the Bavarian light horseman's drill plate.
All photos courtesy of Gilles Cavaillé, Loisirs Detection.

(Left) These Bavarian buttons are made of tombac, an alloy of copper and yellow zinc. The larger buttons on the top row indicated rank of warrant officers. The second and third row buttons were used to secure the soldier's uniform. The button on the bottom with the numeral "1" indicates the company number.

(Above) The French detectorist who located these German one pfennigs arranged them on a brass plate for this photo. Each of the recovered coins pre-dates 1914.

(Above) Bavarian cap badges. The arms of each cross contain the phrase "In Treue Fest," which means "firmly in fidelity."

(Above) American World War I uniform buttons recovered by European metal detectorists. On the top row, the central button secured the pocket of a jacket. The four infantry buttons on the top row, 17 mm in diameter, were used for closing cartridge pouches. The second row of buttons (23 mm size) were used to close an American jacket and the bottom three buttons (30 mm in size) secured an American coat. Trademarks on the backs of these buttons indicate that they were manufactured in 1910 in Newark, New Jersey, and in Philadelphia, Pennsylvania, in the United States.

(Above) German identification plate for Battalion J, a substitute battalion of the regiment of infantry.

(Left) These tombac buttons range in size from 25 to 30 mm. These are buttons of rank from the Prussian army. The larger bottom button decorated the collar of a Prussian warrant officer's uniform.

(Right) These recovered dogtags show the soldier's name, date of birth, and regiment.
(Below) Bavarian soldiers seen inspecting their clothing for lice as a warrant officer monitors them. Note the dogtags around their necks in this early photo.

Many other personal effects items can be found in Europe where World War I troops fought or were stationed.

(Right) These shaving kit items include a razor and shaving goods manufactured in Berlin, Hamburg and New York, among others.
(Below, left) German harmonicas recovered with metal detectors.
(Below, right) Tombac uniform hooks—some removable and some fixed within the uniform.

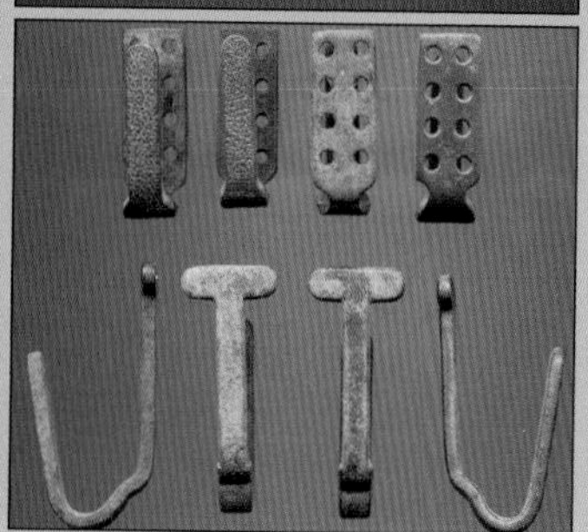

Photos on both pages are courtesy of Gilles Cavaillé, Loisirs Detection.

Italian Detectorist Solves World War II Loss for U.S. Family

A short but touching ceremony in 2004 helped bring closure to a family who had lost their loved one in Europe 60 years earlier. Italian metal detectorist Renzo Grandi, from the community of Borgo Tossignano 30 km southeast of Bologna, had excavated the missing pieces to the puzzle with his metal detector in 1997.

Renzo, a collaborator of the Museum of the Battle of Castel del Rio, happened to make a significant find during one of his Saturday morning hunts. He was searching for the "umpteenth time" on the southern side of Monte Battaglia, the sad side of much bloodshed between Germans and Americans in September 1944. On this steep slope, his Garrett *Gold Stinger* gave a clear signal, one that he immediately recognized as different from the sound made by bullets.

Renzo dug a U.S. Army helmet with a jagged rip through it where a shell had obviously fatally wounded the soldier wearing the steel hat. Renzo decided to return the next week to search that area more closely to see

This German helmet *(right)* and this American helmet were recovered by Italian detectorist Renzo Grandi with a Garrett *Gold Stinger* detector. The helmet at right is that of U.S. Army Private Harry Castilloux, who had been listed as MIA for decades. The items were found west of the Valmaggiore Castel del Rio, where in September 1944 U.S. Army forces had battled German troops defending Monte Acuto. *Photos and story details courtesy of Stefano Morsiani of Electronics Company Italy and Valerio Calderoni.*

if this fallen man had lost any other items. Sure enough, his *Gold Stinger* signaled other targets within 10 to 30 centimeters of where he had dug the helmet. Renzo excavated a bracelet, rings, uniform buttons, some bones and, finally, the soldier's "dog tags" or identification plates.

These were the remains of Private Harry Castilloux of Detroit, Michigan, who lost his life on the night of October 4, 1944, during a German counterattack. A heavy mortar and artillery attack killed some U.S. Army troops on this hill, some apparently buried by earth and rock landslides in their protective trenches. Castilloux's company had moved out from Monte Battaglia the next day and he was officially declared missing in action.

Renzo Grandi's detection work allowed this soldier's remains to be handed over to the American consul in Italy. The bones of the lost hero were finally buried in Michigan with full military honors more than half a century after Castilloux's death.

In October 2004, a ceremony was held on Monte Battaglia to unveil a new memorial stone dedicated to Private Harry Castilloux and the story of his discovery. In addition to the interested historians present, two of the guests of honor were Harry's sisters, 75-year-old Rita and 82-year-old Katie. Harry's commemorative marker stands on the southern side of the hill just a few dozen meters away from where the remnants of the castle's stone tower still stands.

(Left) Detectorist Renzo Grandi, far left, with his metal detecting companions Stefano Morsiani and Scardovi Andrea. *Photos courtesy of Stefano Morsiani of Electronics Company Italy.*

(Above) The helmet and dog tag of Private Castilloux, with the Castel del Rio in the background.

Chapter 10

Bells, Pins, Odds And Ends

Artefacts are of great interest to metal detectorists because of the Old World history that they bring to life after centuries or thousands of years in the ground. Even seemingly simple household items from civilizations long gone take on special meaning.

Viktor Tobolsky of Russia brought his brand-new Garrett *ACE 250* along on a hunt a goose-hunting trip to the distant northern mountains near the White Sea. He and his buddies stayed at the farm of their friend Yuri who had recently found an old cache of coins that he plowed up in his gardens.

"Here was my chance to find some good treasure," Viktor said. "I decided that if a pot of coins was in this garden there was a chance other good things might be hidden in the earth nearby. After a long journey, we arrived at Yuri's home near the mountains. The next day, my friends took their guns and went hunting geese. I was left behind with the duty of making food this day. As soon as the hunters departed, I took my *ACE 250* and went out to search by the vegetable garden. The garden stretched from behind the old Viennese house almost to the forest 100 meters away.

"Almost immediately, I had a signal. Using my infantry shovel, I dug up an old sconce. Five steps forward, I had another signal. This time, I dug up a heavy soldier's buckle that still had the remainder of the rotted belt. It was a brass buckle apparently from before the war. The next signal I dug was a copper coin dated 1785. It was green-colored but well preserved. Within one meter of this coin, I found two other coins, a one-kopeck from 1899 and a two-kopeck from 1901. At this point, I had to interrupt my search to make dinner before the hunters returned.

"Several days later, when I remained behind again as the designated cook, I was able to hunt around the garden again with my *ACE 250*. I found the remains of an old foundation and hunted around it. In these far northern settlements of Russia, there is very little metallic garbage. If a signal is heard, it is usually worth digging. I found small objects to depths of 25 to 30 centimeters. Near the house foundation, there was quite a bit of iron trash, such as nails and wire."

Viktor made the most of his available time to hunt around the farmer's gardens. Many of the objects he found had been crafted

(Below) Detectorist Viktor Tobolsky searches through an overgrown area near his friend's gardens in northern Russia. He managed to find some early Russian coins and also some interesting artifacts of earlier life in his country. Some of the implements below had been forged by a local metal smith, including this miner's lamp, a pick head and a fishing stake.

long ago at a metal smith's shop. Some had been used for mining work in the mountains. "The most interesting thing I found here was a miner's kerosene lamp in very good condition. In almost the same place I found an ancient forged pick near the miner's lamp." In the end, Viktor did not discover another coin hoard but he came away with a new appreciation of some of the area's history.

Metal items have been crafted for thousands of years through bending, cutting, hammering, casting, stamping, pressing, forging, rolling, turning and pressing methods. Veteran detectorists can begin to put an approximate age on an artifact based on its metal type and by its crafting method. Coins, jewelry, fibulas, and military finds have special meaning to some, but the most basic of home supplies can be equally treasured.

Roman wine sieves have been unearthed which date back to the time of Christ. Ancient surgical equipment and pharmaceutical supplies from the medieval period give new insight into the marvels of more modern technology. Some of the ancient hairpins used by European women look almost like small daggers. Bronze Roman and Saxon pins often show fine workmanship, with some containing pin heads decorated with figurines or faces. Early tweezers often contained a small metal scoop on the opposite end which was used to scoop out ear wax. Ancient Roman scalpels from nearly 2,000 years ago have been recovered which look surprisingly little different from the shape of modern surgical tools.

Pipe cleaners from the 17th century were often made of brass and ornate tobacco tins are a great find; some recovered European copper tins date back to the early 1600s. As tobacco smoking became more popular, some of the wealthier citizens began carrying expensive silver tobacco tins during the 18th century. Great craftsmanship also went into decorative tea pots in the more recent centuries. By the 18th century, some were made of pewter while more lavish silver-plated teapots can be found.

Detectorists also find locks and keys of all shapes and size, some dating back to the Roman era. The keys range from the larger ones used to lock heavy church doors to small casket keys and chest keys. Bronze keys are more common to discover because ancient iron keys tend to deteriorate in the soil.

Crotal bells—also known as jingle bells, sleigh bells, rattles or animal bells—are common discoveries on farmland in the UK and throughout Europe. Example of such harness decorations have been found in England that date to as early as the 13th century. Many of the post-medieval bells were constructed of an alloy of copper and tin. Some crotal bells even carry on their underside a maker's mark, a symbol or the initials of the maker.

Countless other everyday "odds and ends" are unearthed by hobbyists each day. For some, these metallic artifacts may be considered rubbish but to others the ancient household items are treasured just as much as a valuable coin. The images displayed in this chapter illustrate just a few of the implements and accessories of ancient life that the detectorist may encounter.

(Above) These post-medieval period crotal bells were found in England by Nigel Ingram. The larger intact bell at left is 4.5 cm in diameter. The small inset photo shows how a group of such bells could be used on a horse's sleigh.

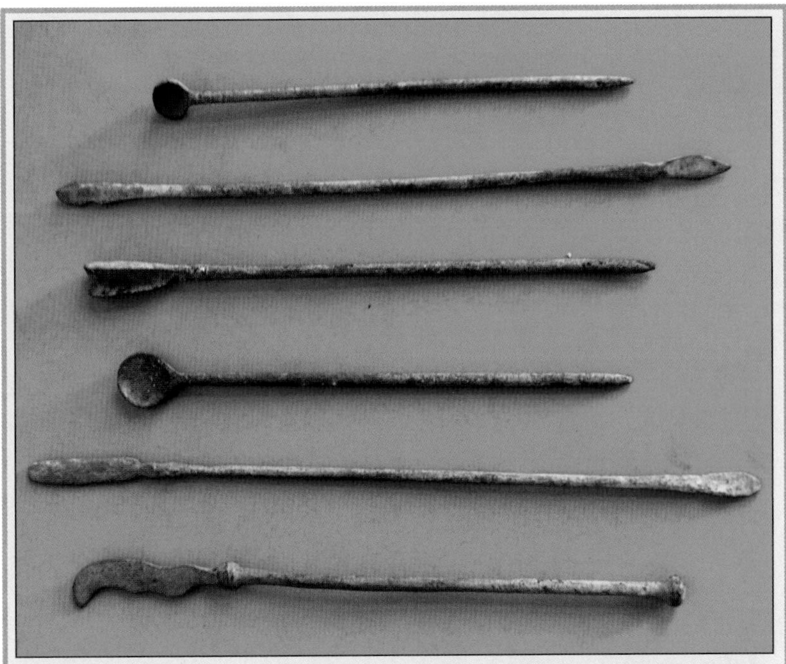

(Above) These ancient bronze surgical tools were used by the Romans around the 1st century AD for cutting, scooping and measuring. *Courtesy of Bruno Lallin, Lutéce Détection.*

(Left) These three Roman bronze chastity belt keys and *(below)* a Roman bronze spoon, about 15 cm in length, were found in a French field. *Courtesy of Gilles Cavaillé, Loisirs Detection.*

(Left) This fragment of a Roman bronze key was found by Regis Najac of France.

(Left) This tiny brass pot was found in Belgium by a Garrett detectorist at a site that was once occupied by Romans and Celts.

(Below) European farmland can yield all sorts of interesting treasures, such as this small bell dug by a French detectorist at a rally.

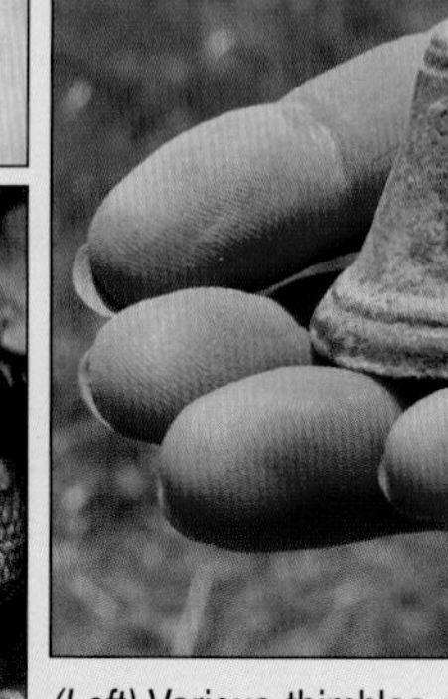

(Left) Various thimbles recovered by Franco Berlingieri.

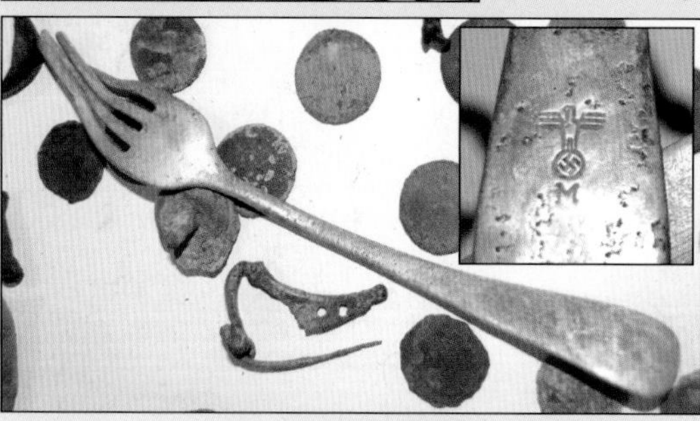

(Left) A piece of Nazi silverware found by Franco *(see inset)*.

(Lower left) A montage of relics all found by Scardovi Andrea in Italy with *GTP 1350* and *GTI 2500* detectors. From the upper left corner moving clockwise, these bronze items are: a Roman key; a spatula; open-looped fibula; bronze cross buckle; another, more ornate Roman buckle from the Byzantine period; and a statuette of Mercury (Hermes the Greek). *Images courtesy of Scardovi Andrea and Stefano Morsiani of Electronics Company Italy.*

CHAPTER 11

SAND AND SURF HUNTING

There is no wrong time of year to visit the beach to search the surf or sand for coastal treasures. Walking the beach with a metal detector is great exercise any time of year. In the spring and summer, you can enjoy the gentle winds off the water as you walk barefoot through the soft sand near the water's edge. You might wear a little heavier clothing at the coast in the colder months, but you can certainly still find great treasures—particularly after a winter storm has churned up the beach.

The interesting and often valuable items you can find along European beaches and in the surf include coins, rings, necklaces, chains, bracelets, anklets, religious medallions, crucifixes and knives. You might also discover coins from an ancient shipwreck, caches hidden by Viking explorers and various other treasures that could date back hundreds or even thousands of years.

The outer European countries are, of course, surrounded by various bodies of water including the Atlantic Ocean, the Mediterranean Sea, the Adriatic Sea, the North Sea and the Baltic Sea, among others. The conditions detectorists experience in various coastal areas of Europe will therefore differ. Beaches along the Baltic Sea and the Mediterranean, for example, have little tidal movement, so discussions in this chapter regarding high and low tide searching will be more relevant for some areas than others. Aside from tidal differences, most of the other metal detecting tips presented here are valid anywhere.

Ancient treasures can be found with good research and a little luck but thousands of detectorists simply enjoy beachcombing on the more popular tourist beaches where their time can be productive. It is easier to fill your treasure pouch with Euros, modern jewelry and various lost trinkets on a busy beach than it is to locate really old items in more secluded areas. There is plenty of "old" European money to be found on the beach since 27 countries have swapped their old currency for Euros. The southernmost European beaches with longer warm seasons equate to greater numbers of total tourists in a season—and therefore more lost treasures.

Loamy or rocky beaches afford sunbathers and swimmers a pretty fair chance of retrieving their misplaced items. The same coins and jewelry dropped on a nice, sandy beach are more quickly swallowed up by the sand. Contemporary coins and jewelry can be found on almost any sandy beach where the public has spent time. Swimmers often forget to remove their valuable heirlooms and diamond rings. These rings expand in the heat, fingers wrinkle and shrivel in the water, and suntan oils merely hasten the inevitable loss. Other rings are lost and necklace clasps are snapped while people play ball, throw frisbees or rough-house on the beach. These lost valuables are thus deposited into the sand and surf. The truly magnificent thing about coastal treasures is that the supply of coins and jewelry is constantly being replenished both by new tourists and the incoming tide.

More than two-thirds of the earth's total surface area—nearly 200 million square miles—is water. Since the earliest days of European history, the various peoples of this continent have lived, played and conducted trade on or near water. Transportation, commerce, recreation, exploration, warfare and the search for food have compelled men and women to return to water time after time. Whenever man made contact with water, he generally brought along valuable items or tools, some of which were inevitably lost. The lakes, streams and oceans in and around Europe thus offer vast storehouses of lost wealth that await the modern metal detectorist.

Henry Tellez searches near the waterline on a Barcelona beach. He is careful to swing his VLF detector's searchcoil perpendicular to the water to avoid passing from wet salt sand to dry sand.

Detector Basics at the Beach

It is important that you select the right detector for where you will search. **VLF (Very Low Frequency)** detectors generally operate on frequencies which range from 3 kHz to 30 kHz. Such VLF machines will do well in dry sand areas and many offer a graphic target display system to help searchers distinguish between potential targets. Be sure to revisit the discrimination illustration in Chapter 2, as the pesky pull tabs so often found on the beach that you might opt to discriminate out may cause you to miss gold rings. Be advised that you will experience some interference in ocean water or on wet salt sand with VLF detectors that do not offer a Salt Elimination mode. Factory settings on most detectors will include some discrimination when you turn them on, but you can regulate this to find all metal targets or to eliminate any that you choose.

Pulse Induction (PI) are ideal for use in the surf and on wet sand—areas where the high salt mineralisation can challenge some VLF detectors.

Swing your detector parallel to the sea as you approach the water. Some detectors, particularly VLF models, will loose their ground balance as their electronics are passed from wet sand to dry sand and back to wet sand.

You must also take special care to keep your detector's control housing dry and out of the water unless, of course you are using a submersible unit such as Garrett's *Infinium LS*™ or *Sea Hunter Mark II*™ models. Deep surf hunters and wreck divers are best served by one of these two models, which are fully submersible to 200-foot depths. Both models use PI (Pulse Induction) technology, which transmits a pulsing signal that is impervious to the effects of mineralization. PI detectors are

This detectorist is taking advantage of the early morning hours to hunt the productive "towel line" with a VLF machine before sunbathers turn out en masse for the day.

thus ideal for use in the ocean for diving to deep shipwrecks or for use in the wet salt sand on ocean beaches.

The more expensive *Infinium LS* model includes features that are beneficial to the detectorist who seriously concentrates on the beach. Underwater PI detectors—excellent for eliminating wetted salt, hot rocks and thin foil—are unable to discriminate the iron targets which can litter certain coastal areas. The *Infinium*, however, offers a solution to this PI challenge. The serious beachcomber can hunt in the All Metal Mode. When a good target is encountered, double-check it by turning the discrimination knob fully to the right. If the signal remains, you have an iron item. If the signal disappears, it is a target worth digging.

Regardless of what brand of detector you select for beach hunting, you should carefully study the instruction manual or DVD that accompanied your unit. This will help you to better understand the pros and cons of any discrimination modes, such as Salt Elimination, that might be available on your detector.

Utilize a methodical pace while searching with your detector in a pre planned pattern. Don't make the mistake of becoming too excited and racing across the sand. Reduce your scan speed to about one foot per second as you let the searchcoil just skim above the sands. Make sure that you keep the coil level to the ground at the end of your swings. Overlap each sweep by advancing your searchcoil about one-half of its diameter and always scan in a straight line. This improves your ability to maintain correct and uniform searchcoil height, helps eliminate the "upswing" at the end of each sweep and improves your ability to overlap in a uniform manner, thus minimizing skips.

Your preparation is just as important as your technique. As you plan for your next beachcombing expedition, prepare a checklist of all the items you might need. In addition to your detector gear, recovery tools, food, drinks, sunscreen and other personal items, make sure to pack extra batteries for your detector(s). You should also include towels and dry clothes in case you are caught in a shower or take an unexpected dip under the waves.

You may not find great treasures every time you visit the beach but here are some general tips that should at least increase your odds of having a productive beach-hunting day:

- **Use headphones while hunting at the beach.** The practice of using headphone is solid advice regardless of whether you are hunting near the water or not. You will simply hear signals better with headphones and you will hear more of them. Crowd, wind and surf noises at the beach can negatively affect your ability to pick up target signals.

- **Use the proper recovery tools** based on whether you are primarily searching dry sand, wet sand, surf or a combination of all of the above. (Chapter 3 includes illustrations of good treasure recovery tools for the beach.)

- **Do not ignore any detector signals.** Always determine their source. If a loud signal seems to come from a can or other large rubbish object, remove it and scan the spot again. When you hear a very faint signal, scoop out some sand to get your searchcoil closer to the target and then scan again. If the signal has disappeared, scan the sand you scooped out. You may have detected a

A good pinpointer can greatly speed your target recovery time and help locate tiny, secondary targets in the sand.

very small object. It might be an insignificant item such as a fishing weight, but you will at least know what caused the signal.

- **Research the area you are visiting.** Find where events have been held on the beach or where the more popular tourist destinations are; people will have certainly lost valuables there.
- On the dry sand, **look for the remnants of beach barbecues or where concessions stands** have been set up. Anywhere that people reached into their pockets for money, you can bet that items were lost.
- **Lifeguard stands** are a good indication of areas where high concentrations of people will have sunbathed and swam on good weather days.
- **Piers and jetties** can be hot spots where people strolled, fished or played. Be careful as you dig in these areas, as they can also be littered with broken glass or fish hooks. Watch for the direction of the tide's ebb to see where lost items would naturally be deposited against large rocks, sea walls, and wind breaks.
- **One of the most productive areas is the "towel line,"** or strip on the beach just above the high tide line where most sunbathers will spread out their towels, blankets and umbrellas. The sand here is generally pretty easy to dig and can yield many worthy targets: jewelry, coins and all sorts of lost items.
- **Search in deeper water for good results in the Mediterranean and on other beaches where there is less tidal effect.** You can often find quality coins and nice gold jewelry by wading out into less-often-searched knee-deep to waist-deep water.

• **A quality pinpointer can increase both the speed and number of treasure targets you recover at the beach.** Garrett's *Pro-Pointer* is water resistant and includes a scraping blade for raking through large piles of sand to pinpoint the more elusive, small coins or gold earrings. Sand scoops work well for many targets but detectorists using large coils for deep depth often switch from the scoops to traditional spades to get to deeper targets. Your pinpointer will be of great value in these situations.

• In general, it is easier to hunt during times when crowds are not present. During peak seasons, **hit the beach at first light or late in the day** to take advantage of the empty seashores. You will also avoid unwanted curious "visitors" following you around.

• **Dress for the conditions you expect to encounter.** During the cooler months, add an extra layer of clothing to protect yourself against a strong, cold wind. You can always remove a layer if you become too warm.

In warmer months, protect your skin with adequate sunscreen and comfortable yet protective clothing. Long sleeve shirts made of porous material can protect your skin while venting some of the heat and perspiration you will generate. Soft cotton gloves are worth consideration during any season to protect your water-soaked fingers against cuts and scratches as you dig out items.

• Finally, **fill in the holes that you dig on the beach**. Don't leave unsightly holes that could cause people to trip. As with any other metal detecting location, **haul out all of the trash that you recover**. You'll be doing future detectorists and sunbathers a favor by removing dangerous metallic rubbish.

Take Advantage of Tides and Treasure Traps

By studying the tides, you will learn that certain times of the day are better than others for hunting the coastline. Nearly twice a day a full tide cycle occurs—two high and two low tides. Low tides are of greatest interest to you because the water level has dropped, leaving more beach area exposed. A one-half foot drop in tide level can expose an extra ten or more feet of ground distance to the water's edge, allowing you to work not only more dry land but also a greater distance into the surf.

Low tides occur about every 12.5 hours. Plan your work period to begin at least two or three hours before low tide and continue long after designated low tide times. That's four to six hours of improved hunting. Tide tables are available online, in newspapers or from local scuba shops and fishing tackle stores.

If you plan to work inlet, cove and river areas, water current data may also be of interest. Take advantage of days—especially after a new or full moon—when there will be lower-than-usual tides. Also, listen to weather forecasts to learn of prevailing wind data. Strong offshore (outgoing) winds will aid in lowering the water level and tend to reduce breaker size and force. Offshore winds also seem to spread out (thin) sand at the water's edge. This effect could result in decreasing the amount of sand that has built up over lost treasure. Conversely, incoming waves and resulting

Any permanent structure or large natural object becomes a natural "treasure trap" that can often be a productive hunting area.

larger breakers tend to pile sand up, causing beaches to thicken and increase the depth of lost items.

Be alert to the lowest or ebb tides when you can work beach areas not normally exposed. You must get your timing right. Of course, you can work dry beaches during high tides and then be prepared to follow the tide out. That procedure offers maximum work time. You must exercise extreme caution if you intend to work with the tidal range's dead low water period to hunt a sandy bay surrounded by high cliffs. Some prime hunting shores can be exposed during the period of low water in such horseshoe bays. Keep careful track of your time to avoid being caught by the incoming flood tide. Since the total time between when the tide begins to ebb and when it returns to its original position is slightly less than three hours, extreme caution must be exercised.

As you follow the tide out, work in a parallel path hugging the water's edge. If your path length is not too long, each return path will be nearly parallel to the preceding one. If your path length is long, each succeeding path will veer outward. Wide searchcoil sweeps can offset these veering paths, however. Be alert to the relationships between locations of your finds. You may discover a trough, or other treasure deposit areas that need additional scanning or work with a larger, deeper-seeking searchcoil.

Look for tidal pools and long, water-filled depressions. Any beach areas that hold water should be investigated since these low spots put you closer to treasure. As the tide recedes, watch for streams draining back into the ocean. These "mark" the location of low areas. If you will constantly keep in your mind the vision that only a few feet beneath the sand's surface a blanket of treasure awaits, your powers of observation will keep you alert to specific areas to search. Continually watch for those low areas that put your searchcoil closer to treasure.

Weather is a major contributing factor to tide levels, and strong storms, and winds can change tides drastically. A storm at sea moving in your direction may raise normal tide levels several feet. When this occurs, wave action becomes so violent it is sometimes

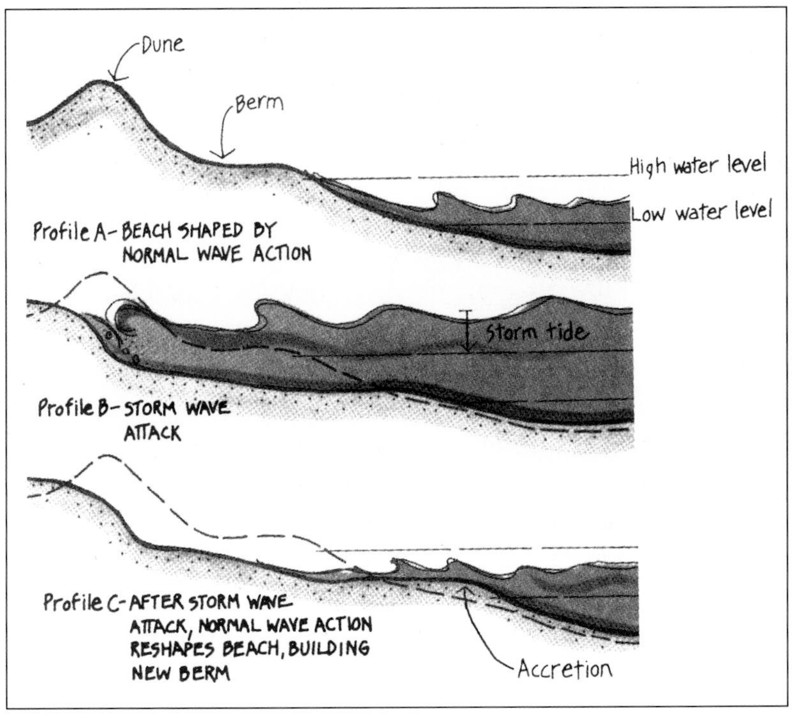

Storm tides can completely reshape sand dunes on beaches, as shown in this illustration from a booklet produced by the U.S. Army Corps of Engineers.

impossible (and dangerous) to hunt, even upon the beach. But, the stage is set and you should hit the beach when calm returns.

Conversely, a winter storm reaching the coast with any strength at all can cause lower tides than those listed in the table and an accompanying compression of wave heights is noticeable. The water is often calm. These conditions and the changes they cause represent a continuing process that controls sand deposits on the beach and in the shallow water. Storms often transfer treasure from deep water vaults to more shallow locations. For a change in your searching habits, plan a beach search right after a storm.

Keep in mind that extremes in weather and surf conditions can make unproductive beaches suddenly become productive. Severe storms with raging seas that destroy sand dunes often create an ideal search environment for metal detectors. **Old coins and**

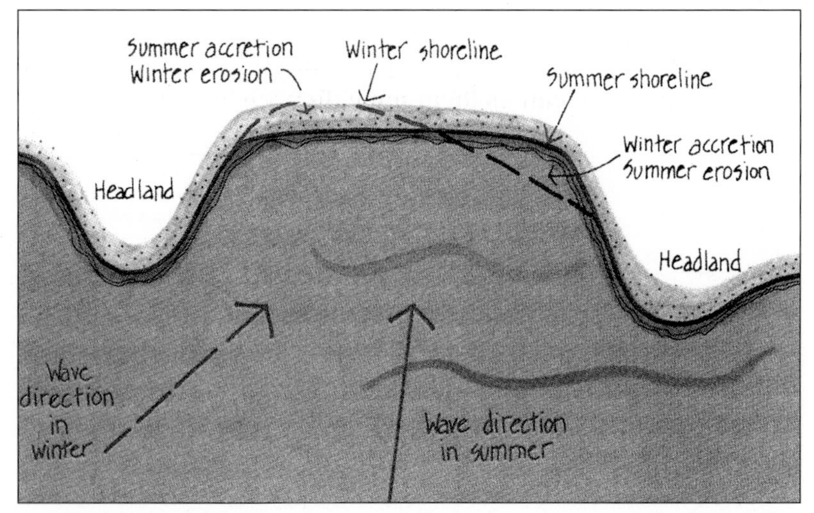

Because beaches continually reshape and protect themselves each season by sands shifting to expose the least possible shoreline to the sea's onslaught, their contours will change with the winds and waves.

other goodies once buried under yards of sand may suddenly be within reach for the first time in centuries. Such heavier metallic items are generally not swept away. Fast-running beach drainage currents can wash deep treasure gullies in the sand. So, keep your eyes on water movement during such violent weather.

Both wind and water move beach sand around in a continual process. This process creates **nature's traps** that will hold treasure for you. Another reason for working beaches immediately after a storm is that the beach continually reshapes and protects itself. Sands shift normally to straighten the beach front and present the least possible shoreline to the sea's continuous onslaught. During storms, beach levels decrease as sand washes out to form underwater bars which blunt the destructive force of oncoming waves. Following the storm, the smaller waves return the sand to the beach.

To understand how sand, coins and jewelry continually move around, observe the relentless action of waves upon sand. At the water's edge, particles of sand form the sand bank. When a wave comes in, the sudden immersion in water causes the grains of sand to "lighten" and become more or less suspended in the water. The constant churning keeps particles afloat until the next wave comes in and carries them some distance by its force.

In the same manner, coins, jewelry, sea shells and debris are continually relocated, generally in the direction of prevailing wind and waves. As they move, waves and wind shift materials about until a spot is reached where the action of the water is lessened. Heavy objects fall out and become concentrated in "nature's traps." So, whenever you find areas with a concentration of sea shells, gravel, flotsam, driftwood and other debris, work them with your metal detector.

Locate areas where the hard-packed sand is exposed. Heavier items such as watches and jewelry will settle in the soft sand down to the area where the beach sand is hard packed. Patches of black sand, where exposed, are good collection areas where the loose sand has washed away. Other areas where the top layer of sand has washed away might be grey, orange or yellow in colour or they might even have a solid rock shelf at the packed layer. You can use a round steel pole to press down into the sand to find out how deep the hard pack is. If the sand is very deep and soft, you can return to this area when the tide is lower.

On your next visit to a beach where surf is especially violent, pay attention. When a wave breaks near the beach, notice that water has a brown appearance caused by suspended sand. When this wave crashes and water rushes up on the beach, it transports sand and mixes it with other loosened beach sand. If the waves break parallel to the beach front, most deposited sand is then washed back into the ocean by the receding water. It remains in suspension in the surf or is deposited near where it came from.

Close observation of the surf will reveal that most waves do not come directly in, but rather at an angle that sets up a current. The sand carried by the wave comes in at the same angle of transport, causing the sand to move farther to the left or right of its origination point. Some of the displaced sand remains on the beach and some is washed back into the water at its new location. **Look for**

Low tide at this beach shows a natural treasure trap. The beach slopes down to this recessed area before climbing up to a sand bar. Work such treasure traps while they are available during low tide and you can clean up on lost treasures.

channels where the water retreats back toward the ocean. Loose items will be swept through these channels and settled down in the sand.

The result of this action is that sand is moved in the general direction the waves are moving. Understanding this is important because this same transport system (via storms and high winds) causes a redistribution of treasure from the point where it was originally lost.

The ability of water to move heavier-than-sand material depends upon its speed. Large waves and fast-moving currents can carry sand, coins, and rings along a continuous path. When wave action slows down, movement slows down or stops. When wave action picks up, movement resumes. "Growing shores" (perhaps those severely eroded by prior storm action) are "nourished" by material that has been washed away from a nearby stretch of beach. Heavy treasure takes the path of least resistance, being pushed up along the lowest points of cuts and other eroded areas.

As coins, rings and other jewelry are brought into these new beach areas, they become fill along with new sand. Being heavier, they gradually sink to lower levels and become covered. When that eroded beach has become fully "nourished," this buildup essentially stops, leaving your treasure buried and waiting for you. It is usually during fall, winter and spring that weather patterns produce major face-lifting on beaches. Strong winds and high tides do most of the redistributing.

As much as 90 percent of the sand on a beach can be washed away during a violent storm. During this erosion process, considerable redistribution of treasure takes place. Unfortunately, some treasure is washed out into the sea, but it still may be found by surf hunters.

Establish your own permanent tide and sand markers to determine how beaches and shorelines are reshaped by the weather. Your marker can be a piling or any structure you can readily observe at any time. Ideally, your water marker will be somewhat submerged especially during high tides. By keeping an eye on this water marker, you can determine water depth at all times and know if the water is rising or falling.

Your sand marker is important because it is a gauge of sand height. The more of your sand marker that is exposed, the greater your chances are of detecting treasure that lies out of reach during those times where sand is being piled up by the winds and waves. The wind direction during a storm is the primary factor in determining whether the sea sweeps the excess sand out to sea or sweeps the sand ashore. Your hunting success will be decreased in the case where excess sand has been swept ashore.

There are high and low sand formations. High formations do you no good except to serve as height gauges when storm and wind activity erode cliffs. Clever beachcombers keep their eyes peeled for cliffs that begin to erode. You are interested in their lowest levels where you can find coins and rings as they become uncovered by the action of winds and waves. Eroding cliffs may reveal decades-old settlements and accumulations of treasure and

Wreck divers using underwater PI detectors continue to recover fabulous European treasures in waters all over the world.

(Above) Two Spanish silver coins from 1733, one as found and one cleaned, recovered by famed underseas salvor Sir Robert Marx.

(Right) Marx surfaces with gold coins recovered from another 1700s-era Spanish wreck.

(Below) Recovered Spanish gold doubloons, also known as escudo coins.

debris. In your research, be alert for references to old settlements or ghost towns. What has been covered for many generations may be uncovered before your eyes today.

Schedule beachcombing expeditions according to current weather reports. Stay alert to hourly weather forecasts (especially for wind chill or incoming storms) and go prepared to withstand the worst.

Those beach hunters who seek the treasures of ancient shipwrecks must obviously spend more time on their research. Some such sites are often found on empty beaches or along rugged coastlines where ships were once torn apart by reefs, rocks and violent storms. Study ancient seafaring maps and accounts of Old World shipwrecks. The most recent book by famed treasure diver Sir Robert Marx and his wife Jenifer—*The World's Richest Wrecks*—provides the last known location of numerous lost ships whose cargoes are now valued at $10 million and more.

Coastal detectorists will dig their fair share of pop tops and assorted metallic rubbish in their quest for treasure. Those hunting the more popular tourist beaches are just as often rewarded with nice rings, jewelry and plenty of pocket change. The truly serious surf and sand searchers—with due diligence and faithful persistence—might just happen upon the find of a lifetime.

Ric Nesbit *(left)* of the *Scuba World* television series examines clumps of coral containing Spanish pieces of eight with Charles Garrett. Using Garrett *Sea Hunters* off the coast of Cartagena, these searchers found numerous such clumps of centuries-old gold coins and cannons.

SAND AND SURF: Tips from a European Detectorist

John Howland, an author and self-described "committed" treasure hunter since 1978, lives near the north shore of Poole Harbor, Dorset. He frequently hunts local tourist beaches, including Bournemouth. John was happy to share some of his beachcombing tips that he has refined over the past three decades.

"Currently, I am using a *Sea Hunter Mark II* pulse induction machine, specifically for beachcombing below the low tide lines of the beaches close to my home on England's south-west coast," John related. "It is capable of recovering coins and rings from over 12 inches deep in seawater-soaked sand without any problem. As I dig every signal, my find rate is more than most others. I would recommend to anyone contemplating buying such a detector to also buy a small shovel or spade, for you will be digging deep targets well beyond what is practical with a scoop or trowel."

In addition to this advice, John volunteered to submit an article on successful beach hunting. In it, he discusses the particular methods he employs with two different types of metal detector technology.

John Howland of England with his arsenal of beach hunting tools. *(Above)* John found this Bulgari 18-carat ring set with 24 diamonds (value $3,650) on the beach with his Garrett *Sea Hunter*.

"Garrett's Keys to the Vault: The *Sea Hunter II* and the *ACE 250*"

by John Howland

The fortune in coinage that lies hidden beneath the sands of the world's beaches just waiting discovery by anyone sufficiently savvy in the art of metal detecting is mind-numbingly colossal.

In the UK alone, official Government figures confirm that between 1983 and 1993 (when the last survey was done) 1,161.6 *million* £1-coins were minted. During that decade, *191 million* of them went AWOL, classified as "wastage," meaning they went out of circulation for any one of a number of reasons. In its report titled *Economic Trends No. 495 January 1995*, the UK Government's Central Statistical Office, attributed the "wastage" thus,

"...they may be dropped in accessible places, taken abroad by foreign tourists, converted into souvenirs, put into permanent collections or lost in a number of other ways."

If just 1/100th of these *191 million* coins were lost on the UK's beaches, and continue piling-up at the same rate to the present day, then, 1.91 million x 26 years (to include the yearly loss rate between 1983 to 1993), confirms that *49,666,000* £1-coins alone, are out there waiting discovery. US Treasury figures will be even more mind-blowing.

Therefore, getting your hands on even a wafer-thin slice of this incredible stash largely depends on using the right kind of metal detector over the ground to which it is best suited, in much the same way, as few self-respecting golfers would handicap themselves by using only one type of golf club.

Where to start then? Firstly, consider Benjamin Franklin's 250-year old political maxim, "*Don't think to hunt two hares with one dog,*" is as good a piece of advice as you will find anywhere. Ideally, you need two *types* of metal detector technology; a low-frequency motion discriminator for the dry sand, and a Pulse Induction (PI) unit for the wet sand.

Hunting Below the High Water Mark (HWM)

Given the fact that seawater holds minerals of all kinds in dilute form, including gold and silver, detecting in or over seawater-soaked sand, is akin to trying to locate valuable targets against a vast sheet metal background—an impossible task. But with the inexorable advance in electronics technology, or 'treasure-onics' if you like, such as those packed into Garrett's *conventional* low-frequency metal detectors, this sheet-metal background can be filtered out allowing access to the coins and jewelry hidden below—well yes, up to a point.

Although the men-in-white-coats at Garrett's "R&D" department are continually pushing out the boundaries of "treasure-onics," even their technically sophisticated conventional discriminators are no match for

These are some of John Howland's most recent beach items recovered in England.

(Left) John found these 2nd and 3rd century coins and fibula brooches with a Garrett *Groundhog* in 1983.

performance of pulse induction systems below the HWM. Without going into the rocket science theoretics of pulse induction technology, except to say that PI's possess an inherent natural advantage in that PI technology is unaffected by seawater mineralization.

Below the HWM, "depth" and the ability to go deep—*very deep*—rules. Garrett's multiple frequency, microprocessor-driven PI detector, the *Sea Hunter Mark II* (I am one of its most ardent devotees), is seldom out-classed. To my mind, this metal detector is the Stradivarius of its kind, which, like its musical counterpart, requires practice and dedication for a virtuoso performance. As South African golfer Gary Player succinctly summed-up his success, "*It's funny, the more I practice, the luckier I become*" then so it is with this beauty.

The *Sea Hunter II* is no opportunist mongrel, but a thoroughbred of impeccable pedigree: The logical development of the groundbreaking and deep-submersible *XL500 Sea Hunter*. This machine was used not only by legendary treasure hunter Mel Fisher to locate the fabulous treasures of the *Atocha*, wrecked in the 1600s off Florida, but also by the distinguished writer, explorer, and treasure diver, Robert Marx.

In 1991, Marx modified an *XL500* to operate at a depth of 1,200-ft, and much of Marx's pioneering research and development manifests itself in the tech-spec of today's *Sea Hunters IIs*. Being fully watertight to depths of 200-ft, these units make unrivalled surf and beach hunting tools. However, be under no illusion, these detectors are pro-quality units, designed by professionals, for professionals.

Sea Hunter IIs will locate coins and rings at depths that users of conventional detectors can only dream: heavy gold rings to over 12 inches, with some coin types, and medallions, to 18 inches, or more. On wet sand, it is peerless. But here's the rub; the pulse induction system is super-sensitive to all things ferrous, making them unsuitable in areas where ferrous junk, prevails. They are quite unsuitable for use above the HWM.

However...below the HWM, where iron presence is rarely a problem and the quality valuables await discovery —where its quality over quantity— this normally debilitating predilection towards iron, paradoxically, becomes the *Sea Hunter II*'s great strength.

Typically, when in the deep-seeking, *Standard Trash Elimination* non-motion mode, (set between "0" and "2") and a signal registers, simply turn the elimination knob clockwise to "9," a position that will eliminate all non-ferrous items. If the signal persists, then the target is iron. If the signal drops away then the target is a non-ferrous, man-made alloy. Simple!...so dig!

Alternatively, small ferrous junk items lying close to the searchcoil, give off loud signals which are easily verified by raising the searchcoil an inch or two; if the signal dies away it's small junk iron. If the signal persists, with the searchcoil between four and five inches from the sand, is usually a dependable indicator to the presence of an alloyed-metal target.

Coins and rings give off sharp, "clipped" audio tones, whereas ferrous targets tend to be brash, with an almost inaudibly "hollow" overtone. Aural discrimination takes time to master. In the meantime, listen, dig every signal, and *learn!*

*Sea Hunter II*s also make other demands. From the outset, users soon realize they are in another ball game, one where targets are deeper than they ever imagined and where traditional sand-scoops and hand-trowels are obsolete. Mini-shovels are the new kids on the block (I got mine from Regton Ltd, brochure code, 2NM).

Above the HWM

Here in the dry sand, ferrous and non-ferrous junk lurks cheek-by-jowl with millions of coins, rings and jewelry. This is where people sunbathe, chill-out, frolic, and play beach games, and it's here where low-frequency, motion-discriminators dominate. In this difficult environment, the Garrett *ACE 250* operating at 6.5 kHz is rarely surpassed (except by another Garrett...maybe!), and is one of the best innovations since bread came sliced and wrapped.

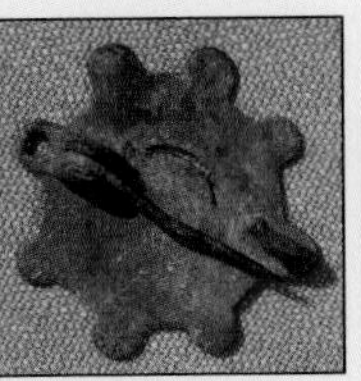

Reverse and front side of a 2nd to 3rd century bronze Roman brooch.

Of its five distinct discrimination modes, the user-programmable *Custom Mode* excels. Begin searching, and manually eliminate the most prolific junk targets as they appear, but err on the side of caution. Some coins and valuables share identical electronic signatures with some junk targets such as pull-tabs, so be wary not throw the baby out with the bathwater.

Nevertheless, if the beach is reasonably junk-free, ease back the Discrim and dig everything. You will be amazed at what ends up in your finds pouch. From one popular section of a local beach, I recovered 497 £1-coins ($705 approx) working two hours a day, (0530 hrs to 0730 hrs) four times a week, over a two-month summer period. All this in addition to 40 £2-coins ($113 approx) and umpteen smaller denomination coins, rings, keys and assorted wristwatches.

Learn the moods of your chosen beach and work it regularly. No matter how thorough your search patterns, you won't find what's not there! Discover first the popular parts of the beach—it's an utter waste conducting a forensic-style search of the less popular areas. Also note the tidal effects on the beach; the direction of the longshore drift; from which parts of the beach coins are prevalent, etc. Once you have unlocked its secrets the beach will be yours and yours alone, forever.

Searchcoils—Large or Small?

Certainly, in this context, size matters, and bigger is not always better either. Both my *Sea Hunter II* and *ACE 250* are equipped with large-diameter ancillary coils, for extra depth, and greater ground coverage. Back-up comes courtesy of a Garrett *Pro-Pointer* that makes child's play of winkling out those smaller, difficult-to-find objects that are not immediately apparent or visible to the naked eye when digging out sand, especially, fishing hooks, or hypodermic needles.

Where high levels of junk exists, smaller diameter coils such as the 4.5" Sniper coil (for the *ACE 250*) are "must have" accessories that will outperform, repeatedly, larger-sized coils. The reason is simple. Imagine two targets under a larger diameter coil; one a pull-tab, the other say, a thin-section gold ring. Pull-tab signals being more robust and dominant, easily overwhelm thin-section rings, thus fooling the detector in to 'seeing' only the pull-tab.

However, the chance of this happening with a small diameter coil, where the probability of both targets being within a 4.5-inch radius is unlikely—though not impossible—the ring would now be safely in your finds pouch. *If you had set the Discrim to reject pull-tabs, then all targets with electronic signatures equal to, or below pull-tabs, would likewise, be rejected.*

This 18th century Catholic medallion is one of John's favorite coastal treasures.

Small diameter coils are also vital tools for searching rocky foreshores, tight corners, and I eagerly wait the day when small coils become available for the *Sea Hunter II*.

And, Finally...

A good and knowledgeable metal detecting retailer is vital—one who not only knows the nuances of his machines, but who is also an active treasure hunter, too. Dealers who know treasure and can talk treasure generally give the very best advice and after-sales service whatever your particular needs.

CHAPTER 12

PROSPECTING AND SPECIAL SEARCH SITES

The preceding chapters have examined some popular metal detecting pastimes including coin shooting, cache hunting and searching along the coast. Just as there are many types of metal detectors on the market, there are many special places that are hunted by today's hobbyists. For gold prospectors, their pursuit of this special metal takes them to the mountains and streams where their odds are better. Other European detectorists have learned to improve their recovery ratios of coins, jewelry and artefacts by working special sites. This chapter will briefly explore some special sites and special metal detecting interests.

Leo Kooistra has used Garrett detectors for more than 20 years of his 28 years of metal detecting. He started out with a C.Scope 990-D metal detector that he, his father and brother first chipped in 960 guilders (equivalent to about $600 U.S. in those days) to buy. On his first hunt, Leo found a gold ring. On his second day out, he found another gold ring. "I was hooked on metal detecting immediately," he admits. "Those things you never forget, finding your first gold."

Since that time, Leo had recovered hundreds of gold rings and thousands of silver rings and pieces of jewelry. His first treasure to find aside from modern items was a 1790 bronze ring. Such early success creates the necessary excitement to become more proficient in the hobby and stay with it for years. One of Leo's favorite places to search for nice jewelry in Europe is in lakes along shores where people swim in the summer. He has found as many as ten gold rings in one day while searching swimming holes. Leo's most productive stint of jewelry hunting was during a camping trip through Germany. Working several swimming hot spots with his Garrett *Sea Hunter* metal detector, he recovered 58 gold rings and more than 200 silver rings in one week.

Gold Prospecting in Europe

Searches for gold in North America, such as the famed 1849 California Gold Rush, attracted prospectors from all over the world to seek their fortunes in the shiny mineral. The latest RAM Books title from Jenifer Marx, *Gold in the Ancient World*, offers incredible insight into the vast history of gold in Europe and the ancient cultures of the world. Gold is mentioned more than 400 times in the Bible. The ancient Greeks and Romans were master craftsmen of gold, and their intricate jewelry, statues, ceramics and ornaments were exported to numerous other countries.

Gold prices set new records in 2009, sparking increased interest in prospecting. It may be surprising to some to learn that gold dust can be found in virtually all European countries with the exception of some flat areas such as Denmark or the Netherlands. The Belgian Ardennes areas was mined in ancient times by the Celts and Romans, and minute amounts of gold dust can still be found in its mountain streams. Gold panning can be practiced by tourists in Finland, Sweden and Norway, and gold panning in Italy dates back to the 1800s. Gold panning is also practiced in certain rivers in Slovakia, the Czech Republic, Switzerland, the United Kingdom, Ireland, Poland, Austria, Germany and France. Gold mines have been opened in recent years in Spain, where Egyptians recovered gold on this country's southeast coast as early as 3600 BC.

Although gold was found in France as early as 2000 BC, professional mining only became serious in the 1900s. Some 1.6 tons of gold was industrially mined in Salsigne in 1972 alone. Tourists still enjoy gold panning in France in four key areas: rivers flowing northward down the Pyrenees Mountains; rivers in the area between Millau, Lyon and Orange; the rivers on the Central Plain; and the lower section of the Loire in Brittany.

Larger gold nuggets, some up to 35 ounces, have been taken in recent years from Switzerland and Scandinavia with the use of metal detectors. The more serious prospectors also take on the challenges of higher, mountainous elevations to search for gold with sluices, picks, pans and crowbars. The largest recorded nugget found in Switzerland was in 1997, and it weighed 123 grams.

In England, gold panning generally results in only small specks of gold but occasional larger nuggets have been found. The largest specimen found in Cornwall is a two-ounce nugget which was located in 1808 during a tin streaming operation. It is now on display in the Royal Cornwall Museum in Truro.

Gold panning is practised in more than a dozen European countries. Adults and even children can enjoy the thrill of trying to recover this shiny mineral.

Gold prospecting has become quite popular again thanks to the sharp increase in gold prices. Special Multi-Frequency detectors *(right)* are ideal for hunting the highly mineralised grounds where gold nuggets are found.

GOLD PANNING INSTRUCTIONS ILLUSTRATED

In this series of photos, Charles Garrett demonstrates proper wet panning techniques with his patented Garrett *Gravity Trap®* gold panning kit. This is the method traditionally associated with panning for gold in streams.

Step 1: Place Classifier atop the 14-inch or 15-inch *Gravity Trap Gold Pan* and fill the classifier with gravel, sand and other materials.

Step 2: Submerge both the pan and classifier atop it, holding firmly with both hands, and use a twisting and shuffling motion to shake them. Small gravel, sand and gold will pass through the classifier and settle in the gold pan.

Step 3: Check for nuggets remaining among the large rocks still in the classifier.

Step 4: Remove the classifier, and discard its remaining contents.

Step 5: Grasp the pan securely with both hands while it is still under water. Begin rotating the contents in the pan as you raise it slightly from the water. Occasionally shake the pan to help cause heavier contents to settle. Remove small rocks as they move to the top of the pan's contents. Occasionally tip the pan forward in the water to permit water to carry off lighter material. Be careful not to lose any of the heavier contents of the pan other than the rocks you remove. Eventually about a handful of concentrates will remain in the pan.

With your hands, break up any mud balls that remain in the pan.

Step 6: Transfer all the material that remains in your pan to the smaller 10-inch finishing pan.

Step 7: Always make certain that the *Gravity Trap* riffles are always on the lower side as you rotate the pan under water. This brings all materials across the traps. You will develop your own method for shaking; i.e., side to side, back and forth or a circular motion. Your aim in moving the pan under water is to cause the heavier gold to settle into the riffles where it will be trapped. As the contents become concentrated on the bottom of the pan and in its riffles, the total amount of material will appear smaller.

Step 8: Continue to tip your gold pan occasionally so that water can carry off lighter materials. Try to separate all other materials from your gold by a gently swirling motion, leaving the gold concentrated together in the riffles.

Step 9: Retrieve your gold. Use tweezers for all large pieces and the gold suction bottle to vacuum up fine gold from the water.

As you follow these instructions, you will develop your own methods of panning. Unusual types of equipment such as square or oblong pans will not help because use of the standard circular pan is natural and is much easier for anyone to master.

You can practice this at home by using small BB shot which will behave like little gold nuggets and settle in the riffles.

The rivers and mountainous areas of Europe where gold can be prospected successfully are generally areas with more highly mineralised grounds. They are thus best searched with Multi-Frequency metal detectors, such as Pulse Induction models. You should start gaining experience with your gold detector by hunting areas where gold is known to have been found before venturing out to new territory.

Familiarize yourself with the various types of ore and rocks present in the area where you will search. Hunt in the All-Metal mode and dig every target until you are familiar with your detector's response to the various minerals in the area. Nugget-sized pieces of gold are challenging to find and require the proper tools of the trade. You will need a sturdy spade and perhaps a crowbar to pry back pieces of rock.

Use a plastic gold pan to help sift through the rocky soil that you dig in response to a pinpointed mineral target. It is obviously important to use a plastic gold pan in order to utilize your detector for checking the contents of your pan. Gold pan kits can be purchased from manufacturers like Garrett that include everything needed for wet or dry panning.

Gold panning is a great weekend activity the entire family can enjoy in scenic locations while camping or on vacation.

Searching for Meteorites

Meteorites—known to many as shooting stars—fall from space to Earth daily, although most are small rocks that burn up as they travel at high speeds through the atmosphere. Meteorites are mainly objects of stone called chondrite (consisting mainly of iron and nickel); achondrites can be as old as 4.5 billion years.

Meteorites are more often found in the ice of Antarctica or in the sandy deserts where they are more easily spotted or detected than on firm ground. Meteorites, many with high iron content, are also best preserved in these terrains. Although in rarer instances, they can also be found in Europe. Dozens of meteorite recoveries have been logged in Germany during the past 200 years, some mere ounces with the largest weighing more than 1.5 tons. Most recoveries of so-called "space rocks" are proven to be false by laboratory examinations, although searchers in Sweden and the Netherlands have unearthed genuine meteorites.

Extraterrestrial rocks have been recovered from a number of areas in the United States. Hector de Luna found five meteorites in 1983 while driving cross country to California. "I stopped at a gift shop in Peachtree, Arizona," he recalled. "The Native American shop owner had a 30-pound meteorite in his shop." The merchant told Hector that he had found the rock in the desert area beyond his shop. "Taking his advice, I started out with my Garrett detector and worked my way from east to west. Within 30 minutes, I had my first meteorite. I started a circular pattern and within four feet of each other, I located another four meteorites." Hector thanked the shop owner and donated one of his five finds to him.

Meteorites can be found all over the world. This 20.6 gram specimen *(above)* crashed to earth on February 12, 1947, in the Sikhote-Alin Mountains of Primorye in Russia. The field from the impact covered half a square mile and the largest crater measured 90 feet by 20 feet. The scientific classification for this meteorite is iron coarsest octahedrite.

This meteorite *(below)* was discovered by Hector de Luna in Arizona in the United States. He had it verified through a laboratory, which found the composition to be 40% nickel and 60% iron.

"Small meteorites exist almost anywhere in Europe," according to Aldino Bartolini of Italy. "Archaeologists hunt for them. Large meteorite strikes can actually be seen on Internet satellite maps. From these overhead views, you can sometimes see the imprints that have been made in the soil in very remote areas."

Other Special Search Sites in Europe

The casual detectorist may be content to search for coins and artefacts in the more common settings—public areas, beaches, fair grounds, along roads and paths, campgrounds, fields, orchards, and vineyards. Others have found their own special places to search. Some of these include:

- *Ridge and furrows*—Hundreds of years ago, many European farmers realized how to make the most of their acreage for farming. A flat acre is simply an acre of tillable soil. Farmers began using their horses and ploughs to pile earth into large mounds or ridges across a field. After considerable effort, many such ridges would be ploughed up in rows across their fields. The resulting ridges and furrows actually created more land for planting and could help with drainage and irrigation.

Today, many farmers have tried to flatten out the ridges and furrows with their tractors but the detectorist with a trained eye can still often spot a field that was once ploughed into ridge and furrow format by the bands of more fertile-coloured crops that persist. In many areas, the acres of ridges and furrows remain as they once were. English detectorists know that such fields were created long ago, often dating to the medieval times, and can be an excellent area to hunt for coins and artefacts of early civilization.

Ridge and furrow fields are often used by farmers as pasture land for their cattle. Unlike ploughed fields, such pasture land can be hunted with permission year round. Detectorists know that once an area is ploughed it is worthy hunting again—even if had previously been considered "hunted out." It is not uncommon at

Medieval farmers often created ridge and furrow fields (*above*) to maximize their farmland. UK detectorists look for such fields as a good sign of early inhabitation.

(*Left*) Pronounced ridge and furrow fields can often be spotted on satellite views taken of farmland. Detectorists use the Internet to help survey potential hunting areas.

all for a medieval coin to be recovered at 20 cm while something much older, such as a Roman artefact, can be unearthed at only 8 cm or less. The effects of ploughing over time can greatly disturb the natural strata where artefacts have been deposited.

- *Ancient Rubbish Dumps*—In the earliest settlements, garbage was often dumped right in the streets of town. Modern cities have been built upon the layers of such rubbish that has settled into the earth. Archaeological excavations have uncovered rubbish dumps from the 1500s which were often located just outside of the early cities. European detectorists should keep an eye out for excavation work and demolition projects. Anywhere that modern progress opens deep layers of soil is worthy of an inspection with your metal detector. Always obtain proper permission from the land owners before scouring such a site.

Such early dumping grounds can produce beautiful coins and a wide collection of buckles, buttons, cookware, thimbles and countless other metallic household items. Other collectors allowed onto such excavation sites have found hand-blown bottles, hand-painted ceramics, tobacco pipes, jars and other non-metallic objects through traditional scouring.

- *Deserted Medieval Villages (DMVs)*—Proper research is necessary to locate such early spots of civilization. Local libraries and archives can contain evidence of an area's earliest settlers. The Internet also contains many postings of DMV sites. You should also look for documented accounts locating the area's earliest markets or fairs. European detectorists who can locate the site of a Medieval fairground can do quite well with finding lost coins and personal items.

DMVs or Roman village sites can often be found in the mountains where they would have been able to view the terrain for long distances. Specifically, Italian and Belgian detectorists have learned to search the southern side of high hills or mountaintops. The southern side of a slope would have helped protect the early inhabitants from the extreme cold of winter because the setting sun would keep the southern side of the slope warmer for longer periods of time.

"You should search on the high side of the southern slopes and then begin working downhill," advises veteran detectorist Albino Bartolini. "Rains and erosion will often wash coins and artefacts down to lower areas over time."

- *Early Roman roads*—The Romans generally built their first roads in close proximity or roughly following natural waterways such as streams or rivers. Romans obviously made camps at various points along their journey and were known to sometimes toss a coin into a river for good luck before attempting a crossing.

Many of today's modern roads were constructed along these ancient thoroughfares. Look for maps or search the Internet for documents which show where the area's ancient roadways once passed through the countryside. Early Roman roads that did not follow a body of water were generally straight as an arrow going north to south or east to west. When you look at modern maps in England and see a road through the countryside that is as straight as if it were drawn with a ruler across a map, you can bet it was once a Roman road.

Some of these roads remain as straight lines. Others county roads built along ancient Roman thoroughfares have been changed from their original course to weave along the more modern property lines that were established in recent centuries. It is therefore possible to study county maps and see where a very straight road occasionally zigs and zags around property lines for a few miles and then returns to its very straight course. With a little imagination, you can almost see that this road was probably once very straight across the land. It might be worth your time to make friends with the local landowners and try to gain permission to search along the line where this old roadway might have crossed their property.

- *Canals and Dikes*—The more adventurous detectorists wade into old rural water canals to search for early coins and other items lost or tossed into the water. In Holland, Leo Kooistra often wades into these canals with an underwater detector and emerges with dozens of coins that often date back hundreds of years.

The first dikes were raised more than 1,000 years ago to protect lives and land from rising seas and rivers in low-lying areas such as the Netherlands. Newer dikes were raised in later centuries as water levels changed, and today many older, unused dikes have become farm roads. Such ancient dikes around old villages or towns can be metal detecting hot spots. Old coins and other valuable artefacts can be found along the sides of some of these remote water barriers.

- *Dredge Piles*—Rivers, canals, community ponds, ditches, moats and other water bodies are periodically dredged to remove excess silt deposits, particularly after a flood. This dredged mate-

Parts of this fortified city's remains date back to the first century AD. This famous restored French castle, Carcassone, overlooks the countryside and distant Pyrenees Mountains. Metal detectorists and archaeologists in Europe have ample opportunities to work together in uncovering history on the sites of early castles and fortified cities.

rial might be dumped miles away or sometimes in a local farmer's field as fill dirt. Many great coins and historic items have been found by alert detectorists who ask permission to scan this dredged earth with their metal detectors. Celtic and Roman travelers often tossed votive offerings into rivers, lakes or streams to appease the gods before attempting to cross the body of water.

• *Rivers, Lakes and Swimming Areas*—Waterways have always served as a gathering place for watering animals, constructing town sites, transporting goods and as campgrounds for expeditions. The fields and banks of rivers can be quite productive, particularly in the areas of an ancient settlement or river crossing. Try these areas during summer months when water levels are lower.

Early bodies of water were also used for bathing and swimming. Changing huts were built adjacent to some lakes and rivers where people gathered to change into or out of their swimsuits. Coins, jewelry and all types of personal items were obviously lost in the waters and around these changing huts.

Charles Garrett had the privilege of detecting on the grounds of the Glamus Castle in Angus, Scotland, in 1986. He had been asked by the Countess of Strathmore to help locate several lost items on the privately owned castle property.

Rivers and streams can shift over time. "I look at maps and visually try to estimate where a river had moved in the past 100 years," says Franco Berlingieri. "Then I try to estimate how far it would have moved in the past 1,000 or 2,000 years." Using his best guesstimate, Franco then begins detecting near the area where the ancient waterway might have flowed.

- *Castles and Fortresses*—Europe was once home to countless fortified homes and castles, and tens of thousands of them still stand. In Germany alone, an estimated 10,000 castle remain. The remnants of such fortified cities and dwellings throughout Europe may never been searched with a metal detector. In most cases, you will need to work in conjunction with the local archaeologists or antiquities societies. Coins, weapons, musket balls, cannon balls, buttons and badges have all been dug with the help of metal detectors around such fortified cities.

These fifteen coins, a medallion and other artefacts from a wide range of history were found by Philip Oyen of Belgium during one afternoon's hunt in Europe. Philip and his metal detecting companions have made several productive hunts in recent years near the site of a deserted medieval village.

CHAPTER 13

RALLIES AND CLUBS

Garrett's marketing and engineering representatives attend European rallies to interact with hobbyists and to learn more about the detecting conditions that are experienced in various terrains. I have attended organized hunt events in America but the European rallies we participated in while researching this book were truly a new experience. Each was different in how it was conducted.

One of the 2009 rallies we attended was held in the spring in southern Spain near Ciudad Real in the fields outside the town of Torreneuva. The early Romans lived in this area long before the time of Christ. Archaeological work in 1934 around Torreneuva turned up Roman coins dating from 268 to 90 BC. Many nonmetallic artefacts have been recovered around this old township that date back as early as the Paleolithic and Neolithic Eras. Such archaeological artefacts can be found in the exact same areas as 2,000-plus-year-old Roman coins and modern Euros lost by farmers in the past few years.

The Spanish rally participants scoured acres of ploughed fields near Torreneuva. Their finds were donated to the town's museum and the mayor was on hand to thank the hunters. The detectorists were rewarded both for recovering specific "planted" tokens and for the volume of garbage metal they removed from the fields.

More than 100 detectorists attended a rally on a private farm outside of Toulouse, France, in June 2009. In this case, the participants again searched for planted tokens and to see what

Metal detectorists literally fill the fields near Torreneva, Spain at a 2009 rally. The competition is good fun, good exercise and offers people the chance to recover for the local museum artefacts that can date back thousands of years in age.

Detectorists at this Spanish rally weigh in the "junk" metal, in addition to the treasures they recovered, for the opportunity to win prizes.

natural artefacts they could find. Rally organizer Gilles Cavaillé awarded metal detectors as the top prizes. Other searchers won Garrett *Pro-Pointers*, digging kits, shirts, hats, headphones and other prizes based on the tokens they recovered. Coins found included Louis XIII *double tournois* coins (minted from about 1610 to 1643), two francs, one sol coin, 28 bronze Napoleon-era coins, two silver Napoleons, a ring, a thimble, a dozen buckles, and various other items, including six ornate bells.

Detectorists will sometimes travel hundreds of kilometers to attend one of these rallies. There is great camaraderie among the

Belgian detectorists await the starting signal at a 2009 rally. *Courtesy of Franco Berlingieri.*

hobbyists who often arrive a day prior to the rally in order to camp near the hunt site. Their early arrival allows them to swap treasure hunting stories around the campfires and compare great discoveries made during the year.

Rallies are one of the best ways to enjoy metal detecting while traveling through Europe. It makes no difference what brand of detector you use or how many years you have been detecting. Some organized events have searches structured to younger age children.

Many rallyists are eager to sweep the field for the prize tokens which have been planted by rally organizers. Others are interested in the opportunity to hunt on private property that has not been previously worked with metal detectors.

Hobbyists often dig up old coins and interesting artefacts that were *not* planted by the organizers. In England, the Wessex Metal Detecting Association hosted a rally on the Englefield estate in September 2007. More than 280 detectorists scoured hundreds of freshly ploughed acres of prime farmland and they managed to find much more than the planted tokens.

The area around the Englefield House has been in continuous use since the Bronze Age. By the end of the rally, the local Field Liaison Officer (FLO) present had recorded more than 100 significant items from Iron Age to modern. The recovered items included Roman brooches, several Saxon pieces, an Iron-Age gold quarter stater from the Atrebates tribe dating to around 100 BC, a Verica quarter gold stater, an Edward III quarter gold noble from the 14th century, and a good selection of medieval and Tudor hammered silver coins. It is not uncommon for such rallies in England to be productive; two detectorists found more than 75 large silver coins during the Newbury Rally of 1997.

Detecting Clubs, Magazines and European Travel

Join a metal detector club in your area to learn more about the hobby and to meet new friends who enjoy the same sport. You will find that it is much more fun to hunt with a group of friends on an evening or weekend than it is by yourself. You even have people around to brag about your good discovery—people that actually know what a *good* discovery is!

Detecting clubs exist all over Europe. In the UK alone, there are numerous regional clubs which can be accessed by searching the Internet. Another source for finding regional contacts is the National Council for Metal Detecting (http://www.ncmd.co.uk). Rarity, a Russian metal detecting club, was established in 1994 and others have since emerged. Belgium, Italy, Germany, Norway, Poland, Sweden and the Netherlands also have organized regional clubs or prospecting groups.

In addition to searching for groups on the Internet, consult your regional metal detecting magazines for more information on groups, organized hunts and rallies. Detectorists post their best discoveries, seek advice from other detectorists to help identify their finds and sometimes compete for annual "best discovery" awards in these periodicals.

METAL DETECTING MAGAZINES

Magazines are distributed in many European countries as the metal detecting hobby continues to gain new members. This list, although not complete, offers an idea of some of the publications that are available. New detectorists might want to pick up back issues to read about techniques used in their area.

England / United Kingdom:

Treasure Hunting—since 1977, monthly
www.treasurehunting.co.uk

The Searcher—since 1985, monthly
www.thesearcher.co.uk

France:

Monnaies & Detections—since 2001, published every two months.

Detection Passion—since 1995, published every two months.

Trésors de L-Histoire—since 1980, published every two months.

Trésors & Détections—since 1982, published every two months.

Le Prospecteur—since 1993, published every two months.

Germany:

Das Schatzsucher Magazin—published at least twice a year.
www.schatzsucher-magazin.de

The Netherlands and Belgium:

The Coinhunter Magazine—published quarterly since 1982.

Detector magazine—published by the De Detector Amateur club for its members.

Poland:

Odkrywca—since 1998
www.odkrywca-online.com

Russia:

Rodnaiya Starina—since 2001, published in Moscow every three months.

Kladez—Annual Russian treasure hunting anthology published each year since 2001.

United States:

These metal detecting magazines each have their share of European readers.

Western & Eastern Treasures—began in 1964 as quarterly *Western Treasures*; became monthly *Western & Eastern Treasures* in 1977.
www.treasurenet.com/westeast

Lost Treasure—since 1966, monthly
www.losttreasure.com

Gold Prospectors—prospecting magazine published every two months by the Gold Prospectors Association of America (GPAA) since 1974
www.goldprospectors.org

American Digger—relic-hunting magazine published every two months since 2004.
www.americandigger.com

Treasure Depot—relic-hunting magazine published every two months since 2008
www.thetreasuredepot.com

Many European countries support monthly or bi-monthly metal detecting magazines that report on the popular sport. Some of the American treasure hunting titles also have subscribers in the European countries.

Our North American friends should not feel left out from this whole European detecting experience. It is possible contact detectorist clubs in other countries before you travel to Europe. Garrett's website now offers a page in its Hobby Division where contact information is shared for those who have an interest.

Speaking from experience, I can say that traveling with your metal detector is not a problem if you are visiting Europe. You will obviously need a current passport, and you should read up on the antiquity laws of the country or countries where you intend to hunt. Your detector can be disassembled and packed into a larger suitcase with clothing packed around it for protection. I know of others who have flown with their detectors disassembled in their carry-on bag who have not experienced problems with security checkpoint screening. I do not, however, recommend carrying a hand-held pinpointer in your carry-on bags. My *Pro-Pointer* was the subject of much scrutiny when an x-ray screener spotted it in a Spanish airport. Such items that might be mistaken for a weapon on an x-ray machine are best stowed in your checked baggage!

As far as digging tools, you will want an army spade or sturdy short-handle shovel. I would recommend contacting the person you are in communication with in Europe about borrowing or renting an appropriate digging tool. As a last resort, you could always purchase an inexpensive shovel at a hardware store upon arrival. If you are attending an organized rally, there are also a number of vendors on site that often sell recovery tools. Another discouraging factor about packing your own folding shovel in a checked bag is the extra weight and the potential overage fees that you might incur.

UK Rally Experiences

The English rally Henry Tellez and I attended in September 2009 was by far the largest in terms of attendance that I have experienced. It was held on a large, private farm west of London in the vicinity of South Oxfordshire at West Hanney. More than 1,000 people turned out to metal detect and help work this major two-day event. Looking over the wide variety of metal detectors carried by the participants, I did some quick math in my head. Guessing that an average purchase price was $400 per detector—multiplied by 1,000 detectorists—there was nearly a half-million dollars worth of metal detectors running around at this site!

People started showing up on the designated farm on Friday to set up their tents, campers and vendor tents. Like those in France and Spain, the rally was a social event where friends, detecting clubs and even families enjoyed cooking around their campfires and swapping stories for many hours.

DETECTOR CLUBS TRAVEL EUROPE

Some European detecting clubs enjoy traveling to other countries on holiday in their quest to find early coins and artefacts. Franco Berlingieri and his "demuntzoeker.be" club from Belgium *(seen above in Italy)* make such annual trips.

They visit farms in Italy where the club is granted permission by the landowners to search on their property. Franco and his Belgian friends have stayed at a bed and breakfast north of Rome during the past few years because the land has been productive. On a high hill on this farm, an ancient village once faced the distant Mediterranean Sea. Aside from productive metal detecting, the club members enjoy sharing good food and taking in the local culture.

Franco Berlingieri and his friend Mark Ickx with some of their finds from a recent trip to private property in Italy.

Saturday was Day 1 of the UK rally. Nine different fields were marked off as hunt areas for this day. The size of each ranged from about ten acres to more than 50 acres. Each had been freshly ploughed to stir up the coins and artefacts below. Groups of metal detectorists stood ready at each of the fields prior to start time. At the given blast of the horns, people swarmed into the fields and hunted eagerly. The finds were good during Day 1 and those who hung around the vendors' tents watched as the first hunters returned to turn in their finds.

Regional British FLOs were on hand to document each find as the people returned. By lunchtime, word had gotten around that Field 7 had a large number of Roman era artefacts and coins that were being turned in. As would be expected, the afternoon crowds packed in Field 7 to try their hand at finding Roman. The other fields were notably reduced in numbers of searchers.

The interesting fact is that any and all of the first day's assigned fields had the potential to turn out good finds. Although most of the Roman was concentrated in Field 7, it was actually Field 1 nearest the vendor area that produced one of the day's best finds. Less than 50 meters from the point where most detectorists entered this field from the road, a brilliant coin was found in the afternoon in an area that likely had been walked over by dozens. It was a gold Celtic stater about 2,100 years old which looked as if it might have been molded yesterday.

I found this rally to be eye-opening in size and a thrill to experience. Detectorists from all over England were friendly, happy to show the coins and artefacts they had discovered, and open to pass on knowledge of areas they had found to be productive.

Day 2 of this UK rally proved to be the most exciting. The official start time was 9:00 a.m. and the detectorists were en route to their chosen field long before that time. Eight new fields were open to hunt. Some opted to head toward the ridge and furrow field but soon found it closed due to an abundance of cattle that had not been cleared. Others made for the field where a deserted medieval village (DMV) had once stood. Still others chose to attack another field through which an early Roman road was rumoured to have once existed.

The air horns blasted like a freight train's warning at a railroad crossing, and hundreds of detectorists again swarmed into the ploughed pastures like ants on a picnic. One of the least likely fields produced at least half a dozen excellent silver staters. As would be expected, people soon began to concentrate in this field to try their luck at finding more staters.

Before lunch, one of the more distant fields became the center of attention. Chris Bayston, a 56-year-old from Yorkshire, picked up a deep signal and began digging down to about a 33 cm (13 inches) depth. In one of his shovels, he unearthed a metallic disc that looked as if it could be from an old tractor wheel. When Bayston brushed the dirt away from the object, he discovered it was far more valuable. He had discovered a piece of Saxon jewelry, a complete brooch made of a copper alloy that was decorated with gold, garnets and coral.

In addition to the rare jewel, Bayston found a skull and other bones of the Saxon who had been buried with the brooch. Rally organizers immediately summoned the local coroner and police, as per British law. Early estimates were that this Saxon grave could well date back to the sixth century, making Bayston's brooch 1500 or more years in age.

Peter Welch, the Weekend Wanderers Detecting Club official who organized the rally, deemed this broach as the biggest find he had seen in more than 20 years. "It could be a Saxon princess or queen," he speculated. Oxfordshire County Council's FLO Anni Byard agreed that this was "an important find" that may have been a royal burial site. Proper archaeological excavation, which has already commenced, will show whether other precious items or human remains are in this area.

The Saxon skull and brooch site became the focus of much police and official activity through the rest of this rally. The brooch, still largely covered in soil, was put on display in a glass case at the FLO table in the big tent. The field where the Saxon treasure was found was quickly roped off like a crime scene until specialists could be reached. Location of such a find becomes a monument area; farming and detecting immediately becomes forbidden within a specified distance. Some early speculations were that finder Chris Bayston and the 49-year-old farm owner might *each* eventually be paid anywhere from £15,000 to 30,000.

As I hiked along the dirt road toward one of the fields at the UK rally, I happened to spot a shiny English five-pence coin and picked it up. As I studied the details of this common currency, I considered for a moment just how often coins have been lost throughout history. Here was a coin obviously dropped by one of the many detectorists which had become another item to be found years from now. I also noticed a couple of young boys who had found several interesting artefacts on the ground near the vendors' tents. People coming in from the fields rummaged through their pockets to purchase items. In the process, they dropped coins and even some of their great discoveries from the field, making them available for someone else to find.

It just goes to show that treasure replenishes itself at a rate that will keep metal detectorists happy for generations to come.

THE RALLY EXPERIENCE

Scenes from a French rally near Toulouse. *(Left)* The whole family enjoys this sport. *(Above)* Detectorists head toward the field where the day's first event is about to start. *(Below)* Men, women and kids all compete to pinpoint and dig treasure targets.

(Top) Detectorists hard at "work." *(Above, left)* Garbage metal is discarded. *(Above, right)* Tokens and good discoveries are saved for the turn-in.

(Below, left) Detectorists return to the turn-in table to have their finds logged in by the rally staff. *(Below, right)* Another happy hobbyist claims his prize, a Garrett *Pro-Pointer*, from rally organizer Gilles Cavaillé.

RUSSIAN RALLY

Throughout Europe, the rally experience is enjoyed equally in many countries. This September 2008 rally was held east of Moscow in Russia. *(Right)* Morning breakfast around the fires before the hunt starts. *(Below)* Searchers await the starting signal.

(Clockwise from above, left) The Russian organizer gives last-minute instructions to the searchers as Garrett representatives Bob Podhrasky, Brent Weaver and Henry Tellez look on. Detectorists begin scouring the field for hidden tokens that can be redeemed for prizes. The searchers are often rewarded with unexpected early-era Russian coins in addition to the items planted by the rally's organizers.

UK RALLY

We attended this large rally west of London near Oxford in September 2009. More than 1,000 people turned out to metal detect, vend and socialize for this two-day event. *(Right)* Detectorists arrived the day prior to set up their camps and enjoy the social life.

(Left) Nigel Ingram (center) and Henry Tellez (right) present one of the donation prizes to the UK rally organizer, Peter Welch of Weekend Wanderers Metal Detecting Club.
(Below) ACE 250 user Paul Hardy of Barnsley and some of his discoveries, including a large token, a flat coin and an interesting horse head ring *(see lower photo below)*. On his first outing with his *ACE* 16 months ago, Paul found a King John silver coin, circa 1200 AD.

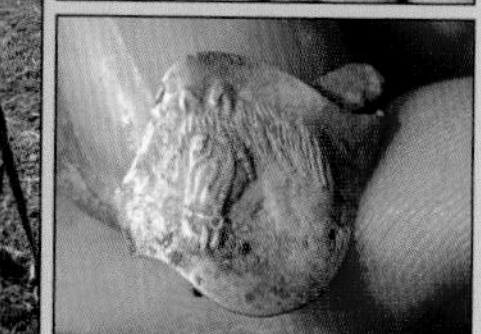

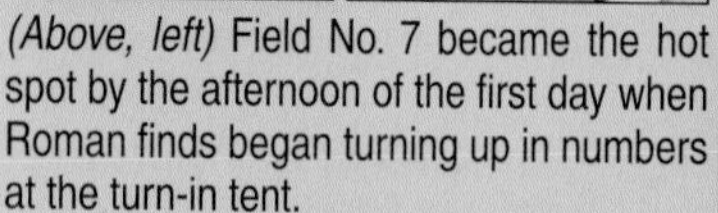

(Above, left) Field No. 7 became the hot spot by the afternoon of the first day when Roman finds began turning up in numbers at the turn-in tent.

(Above, right) The UK FLOs were kept busy recording all of the finds as the turn-ins increased.

(Right) Denise, a British detectorist, looks pleased with her latest recovery made with the help of her *Pro-Pointer*.

(Below) Craig Wall and his 13-year-old son Matthew *(center)* enjoyed attending their second rally together. Craig found this musket ball with his *ACE 250 (left)* while Matthew holds out some of his day's finds *(right)*: a Victorian-era key, pewter buttons and an iron buckle.

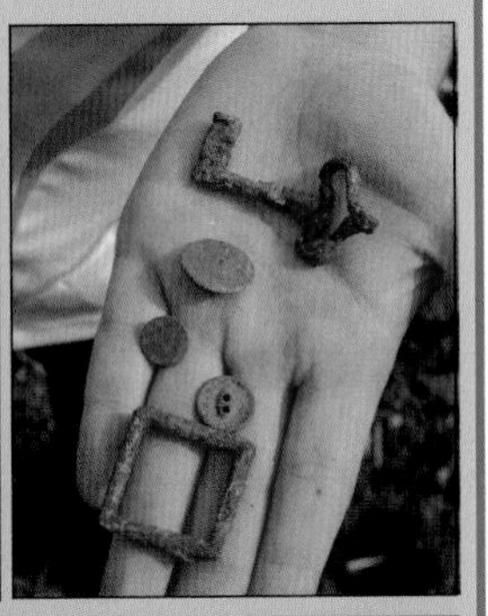

(Above) Detectorists converge on one of the fields as Day 2 prepares to start.

(Right) The landowner moves searchers to the fields with a trailer pulled behind his tractor.

(Below) The rally starter truck sounds its horns at 9:00 a.m.

(Above) Teamwork strategy often pays off. *(Left and below)* The farmer works over one field to create fresh hunting areas. Detectorists waste no time in sweeping the freshly turned earth.

A genuine medieval jewel. This Saxon brooch was found on Day 2 of the rally near Oxford, along with a skull that has been estimated to be as old as 1,500 years. The copper alloy brooch contains gold and is studded with garnets and coral. Some have speculated that the value could be assessed at £50,000 or higher.

CHAPTER 14

CLEANING YOUR TREASURES

Once you have found your European treasures, you will be faced with the decision of *how* or *if* you want to clean away any corrosion or oxidation. *Corrosion* is the primary means by which metals deteriorate as they react to their environment.

Different metals react differently to the effects of the soil or water in which they have been lying until you recover them. Gold is one of the precious metals that generally holds up well to the effects of oxidation. Pure gold coins may be retrieved with only minor discolouration or a thin film visible to the naked eye. Because pure gold is soft, coins were generally created with an alloy of other metals such as silver or copper, which results in further deterioration in the ground.

Pure silver does not oxidize because silver does not combine naturally with oxygen. Silver alloys, however, will corrode and oxidize. Pure copper was often used in coins of the ancient days, but it became a more common alloy in coins of recent eras. Brass is a mixture of copper with 10 to 50 percent zinc mixed into its formula. Three of the most common forms of degradation of copper and its alloys that occur are *tarnish* (usually dull, gray or black in colour), *copper carbonate* (bluish-green weathering of metal) and *verdigris* (a green patina that forms on copper, brass or bronze long exposed to air or salt water).

Copper and copper alloys such as bronze and brass often emerge from the ground with various coloured *patina*, the result of years or centuries of external chemical influences on the surface of the coin or artefact. The thickness of this patina and its colour vary based upon the chemical influences. Dirt, moisture, carbon dioxide, salt content and organic acids all influence the appearance of an old coin. Some coins emerge from the ground with a green patina while others can take on a red or brown layer of corrosion.

The natural tendency of a metal detectorist is to begin rubbing the dirt from an old coin immediately to learn its mint date. Be advised, however, that the most valuable coins can be damaged by even the simple process of rubbing off the soil with your fingers. The value of your coins can best be preserved by leaving the dirt on the coin until you are able to work with them later. It is better to place the dirty coin into a protective pouch or cotton-filled container in your treasure pouch while you are still hunting.

How Should I Clean My Coins and Artefacts?

Most of your finds will be covered to some degree with a layer of dirt which can be rinsed off with water. If you decide to clean your coin or artefact more thoroughly, **you must first decide what type of metal it is**. This chapter is designed to present an overview of some of the available options. The safest process, of course, is to trust your potentially valuable finds to an expert who is experienced in cleaning such items.

If you choose to clean your own coins and artefacts, be aware that it is very easy to ruin them. Yet, you will hear and read about numerous home solutions that people use to clean recovered items. Among the many solutions are: malt vinegar applied with a soft brush; white vinegar or lemon juice; soaking coins in Coca-Cola; and even using red-hot sauce to remove dirt from exterior surfaces. Some may prove successful; others could completely ruin your precious find. *If you have recovered coins that may be of great value, visit a reputable coin dealer or other expert before attempting to clean them yourself.*

Never attempt to remove dirt from a coin's surface with any sharp object that can permanently scratch its surface, such as a knife or a needle. One of the first steps to safely cleaning your recovered coins and artefacts is to soak them in a solution of warm water to help loosen the surface grime. The basic rule of thumb for bronze items with a green patina is to simply not try to clean them. Oxidized silver items can be cleaned with a baking soda or citrus acid solution or through electrolysis. Modern scouring machines should not be used as they will remove patina from objects. The fine detail of older items can quickly be destroyed forever by these more automated processes.

Some coin hunters use a very soft bristled brush with a Pepsodent type toothpaste substance to gently brush away some of the tarnish on old silver coins. Be advised that as you uncover more of the coin's nice surface features, you are also scrubbing away the natural patina that helps give the coin its value!

In extreme cases, scientists or restoration experts opt to employ aggressive cleaning methods to remove corrosion and oxidation. *The use of drain cleaner is not recommended* because of the poisonous nature of the fumes and the fact that this treatment may ruin already damaged coins. Scientists sometimes use this method to strip brass and copper-nickel artefacts with undamaged surfaces.

Others opt to clean coins and silver jewelry recovered from the ocean in an **electric tumbler** used to polish rocks. This method is not suitable for antique items and must be experimented with before you take a chance on damaging an old coin. Another method used by some hobbyists and scientists is **electrolysis**, which uses a vinegar/water solution and electricity to strip items via a chemical reaction.

Experts trained in archaeological restoration have access to more elaborate mechanical and chemical methods and even use x-ray machines. Some restorers may even use a scalpel under a microscope to peel away loose outer layers of patina to improve the appearance of a badly corroded coin or other artefact. Patina removal can quickly ruin the item when attempted by an amateur.

Metal Identification and Cleaning

To properly clean and preserve your metallic artefacts and coins, you must correctly identify the type of metal. Your field experience will increase with your finds and you will be able to better judge an item's composition by the types of patina or rust covering that has formed on it. By carrying a small pocket magnet in your treasure pouch, you can quickly verify if an item contains iron, nickel or steel (an iron alloy). Other pure metal objects (gold, silver, tin, lead, copper, bronze, brass, etc.) are non-magnetic.

The following list is intended to serve for general reference as you learn to identify treasure items for cleaning. In addition, notes on how to clean or whether to clean an item at all are given. Some of the cleaning methods listed under these metals will be more fully described.

• **Bronze**—This metal alloy of copper and up to 10% tin is a hard and heavy substance. Some Bronze Age variations even contained small percentages of lead. Bronze items, which can date back to 2,000 BC, are not pliable and can break under pressure.

Corrosion on bronze artefacts and coins can be in a green, brown or bluish patina. A soft patina layer is a poisonous form of corrosion that should not be disturbed. Experts advise not to clean the patina from antique bronze because it may ruin its value.

• **Silver**—Pure silver coins tarnish easily to a yellowish-brown, gray or black colour. Old silver pieces may include a maker's mark, town mark or a number for its percentage of silver (such as 800 for 800 parts silver to a thousand). Many silver items were alloyed with other metals and can thus tarnish differently. Medieval silver coins with a high copper alloy will turn greenish or brown like bronze. High quality silver remains smooth even if it turns black.

You may also find fairly clean high quality silver in the field because ammonia in a farmer's fertilizer has acted as a cleaning agent on the silver item. The best methods for cleaning silver will depend upon the amount of oxidation present and the quality of the silver alloy. When in doubt, start with the least invasive method *(see next section)* and determine results from there.

- **Gold**—This precious metal had been used for jewelry since the Bronze Age and it remains virtually unchanged in the ground. Gold is not affected by acidity, is heavy and is relatively soft. Gold is therefore generally mixed with other metals to give it strength. Items made of gold alloy may show spots of green (copper mix) or purple (silver mix). Jewelry from the past several hundred years may contain a hallmark to indicate the object's gold percentage. For example, 8-carat gold is shown with the numerals 333, 14-carat as 585 and 18-carat as 750.

- **Brass**—Brass, an alloy of copper and zinc, became more important than bronze during the Middle Ages because brass was easier to work with. Copper is a red-brown metal but brass and bronze are yellow metals that can appear to be gold when dug from the ground. Brass turns dark brown after being in the ground for a long period of time, and it can also take on a green colour when its surface is corroded.

Brass items clean up easier than silver. Start with soap and a soft brush to scrape away the corrosion layer and use lemon juice to rub away any marks. Then, rinse the object and lightly polish it with a soft cloth.

- **Copper**—In its pure form, copper has a red shine. This metal can easily be confused with brass, but copper is softer and thus polishes well. Copper coins often remain in better condition in the soil than coins made of brass or bronze. The ground will eventually make copper coins turn dark brown, and corrosion can turn them black or green in colour. Axes, spears and other tools made using copper have been found that date back thousands of years.

Cleaning copper is similar to cleaning brass. Start with a simple soapy water and soft brush combination.

- **Nickel**—Many modern coins are made of a nickel-copper alloy that can cause them to oxidize to a brownish colour. Any pure nickel artefact, however, will not rust or take on a patina. Nickel is more brittle than silver and is magnetic if it has not been alloyed.

This metal has been used since the 18th century for buttons and buckles, which generally hold up well. Very acidic soil can cause corrosion, however, particularly when the item is a nickel alloy.

Wash nickel items with soap and a stiff brush or use the ammonia cleaning method covered in the next section.

• **Lead**—Lead has been used for many centuries to create bullets, musket balls and weights because of its low melting point. After it has been in the ground for many years, lead begins to turn black and corrode with a white or brown patina as it reacts with the ground acids.

You should not try to remove this coating of patina when cleaning lead artefacts. Simply wash such items with soap and water and use a soft bristle brush to remove soil.

• **Tin**—This soft metal was used since 2000 BC as an alloy when making bronze. It was used by the Romans and Greeks for spoons and other utensils. By the 14th century, tin was heavily alloyed with lead, which produces a harder, darker-coloured artefact. Tin pest or tin disease corrosion creates rough, black porous marks on the surface of items high in tin content.

Medieval tin is very soft and should only be cleaned by using warm water, soap and a soft brush to avoid scratching it. Experts can remove some of the tin pest with the electrolysis method.

• **Aluminum**—This lightweight metal was used by countries such as France, Germany and Greece prior to World War II for some of their lesser denomination coins. Aluminum coins and artefacts will be fairly recent in age (compared with some other finds). Corrosion causes aluminum to take on a yellow or whitish colour as it loses its sheen. Cleaning should be attempted only with soap, water and a soft brush that will not mar the metal. You can preserve the item with a coating of Vaseline.

• **Wrought iron**—Such items are generally preserved well only if they have been dug from mud or dry clay, as soil acids, moisture and oxygen cause most such items to quickly rust. Magnetic wrought iron, used since around 700 BC, will form a thick reddish or brownish crust after being in the soil for a few years.

You can remove some of the exterior crust from a wrought iron artefact by tapping it with something hard. Some detectorist use Coca-Cola (which includes phosphoric acid) as a soaking solution to remove rust. Others use a 5:1 water and citric acid bath and soak artefacts up to 20 days. Once the worst of the rust has been removed, carefully remove and dry the artefact while avoiding contact with your bare hands, which would cause the item to rust further. You can conserve the item by treating it with gun oil.

- **Cast iron**—This tough yet brittle metal was molded into shapes since the 19th century. It corrodes with light to dark brown layers of rust which are uneven. Some loose outer layers can be removed by tapping the item with something hard. Some of this rust can also be removed with a strong steel brush. Citric acid or phosphoric acid can also be used to remove it. Add a layer of Vaseline to cast iron coins after cleaning them.

Common Cleaning Methods Described

The method you use to clean finds varies by metal type. Several methods will be explained that have been used by detectorists to clean up their finds. Refer to the preceding section to learn if any of these methods are suitable for the metal composition of your treasure item. *After completing one of these methods,* your final step should be to neutralize most items by washing them under running water and drying them well.

- **Pure water or soap and water solution**—All metals except iron can be soaked in pure water overnight or for a full 24-hour period to begin loosening deposits. The water is doing its job if it is dirty after several hours of soaking. Use a soft brush to gently loosen other deposits. For the most delicate artefacts, a cotton swab can be rolled carefully over the item to absorb water and dirt.

Add pure soap flakes or soft soap (with no additives) to a tub of warm water. Soak any non-ferrous metal find in this for up to 24 hours and gently brush off excess sediment with a soft brush or cotton swab. Be sure to rinse off all traces of the soapy solution before final drying.

- **Citric Acid Mixture**—Mix citric acid crystals and distilled water in a 1:4 ratio in a container. Pure lemon juice is used by some hobbyists but beware of other additives in the lemon juice that could be harmful. This method is best for silver coins and artefacts with only light tarnishing. Soak items in this solution for several hours or overnight and then use a soft brush or cloth to gently remove the loosened crust. Because citric acid can strip objects down to their bare metal over time, monitor this process closely. Dilute the solution with more water if cleaning seems to be happening too quickly. A citric solution can be used with most metals to help loosen corrosion.
- **Household Ammonia Mixture**—Use a small container to mix a solution of 1:3 ammonia to water for silver coins. Cover the solution and do not inhale the fumes as you let coins soak for about 20 minutes. Use a soft brush to gently remove tarnish and dirt. Repeat this process if necessary.
- **Sulfuric Acid Mixture**—Pour water into a container first and then add sulfuric acid at a ratio of 1:20. Be sure to read the warning labels on the sulfuric acid; avoid breathing it or getting it on your skin or clothing. If you choose this aggressive method, you should soak coins in this solution for no more than ten minutes.

Make sure the coins you soak in this mix do not contain any significant percentage of copper, because the acid will damage copper. This treatment is mainly for removing heavy oxidation on silver artefacts and coins.

- **Silver cleaners**—Various commercial silver cleaning compounds are sold by coin dealers and other retailers. Coins that have become badly discoloured can be cleaned with one of these solutions and cotton balls. Use this solution conservatively, swabbing or soaking the item. Some hobbyists swab down gold items with such a solution as well, but be warned that the chemical can strip some items (particularly copper alloys) down to bare metal if they are not used with care.

(Right) These three Bronze coins illustrate the danger of cleaning old coins too much. The greenish patina has been partially scrubbed from the Antoninus Pius seen at top left (circa 138 to 161 AD). The Seslertius Faustina coin (top right, circa 145 AD) has been partially cleaned. The untouched Hadrian coin (bottom, circa 117 to 138 AD) holds more current value with its original patina. All three coins are about 3 cm. in size, weighing between 23 and 26 grams.

The Kooistra Silver Cleaning Method

Leo Kooistra of the Netherlands was kind enough to share his four-step process for cleaning old silver coins. Depending upon the condition of the coin, he might have to repeat one or more steps in his process to bring the coin to an acceptable condition.

Leo keeps three small pans of liquids for coin cleaning. The first (Step A) is a small pan or tub of "sugar water." For this, he purchases from the pharmacy a soda-type powder known in the Netherlands as "maagzout." This substance is a white, powdery soda-like material that has the texture of granular sugar or salt. He takes a small spoon and mixes two small spoonfuls of this white powder into a small pan with about 16 ounces of tap water. The result is a solution that has the effect of a citrus-based sugar.

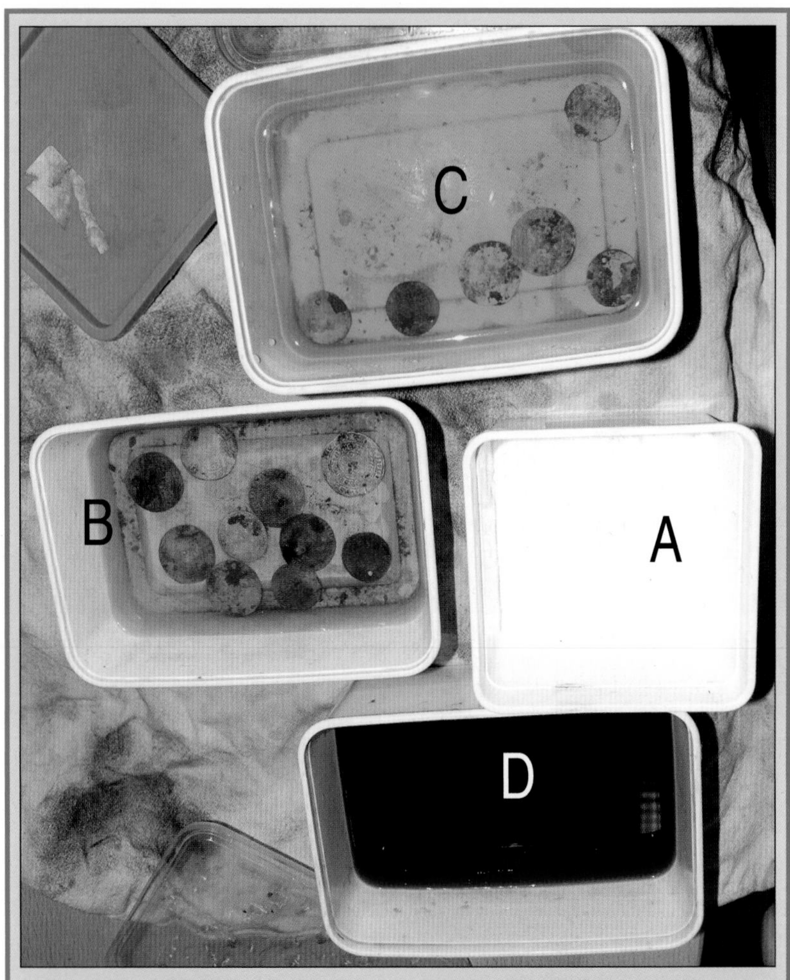

(Above) Leo Kooistra of the Netherlands offered this process for cleaning old silver coins. It uses these four tubs of solutions. Tub A contains the white, powdery citrus sugar he obtains from a local pharmacy. Tub B contains the tap water mixed with the sugar powder. Tub C contains regular distilled water for soaking the coins. Tub D is the water/ammonia mixture.

(Right) Leo holds a sample stack of badly corroded silver European coins that he is cleaning.

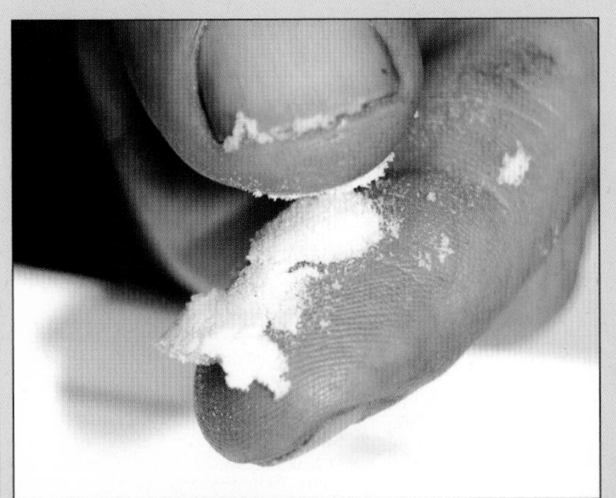

Leo uses this powdery citrus sugar from a pharmacy *(left)* during his cleaning process.

(Below) Leo carefully rubs the powder across an old coin he has removed from its soaking tub. Be careful not to rub the coin too hard as this gentle scouring agent can damage the coin's surface.

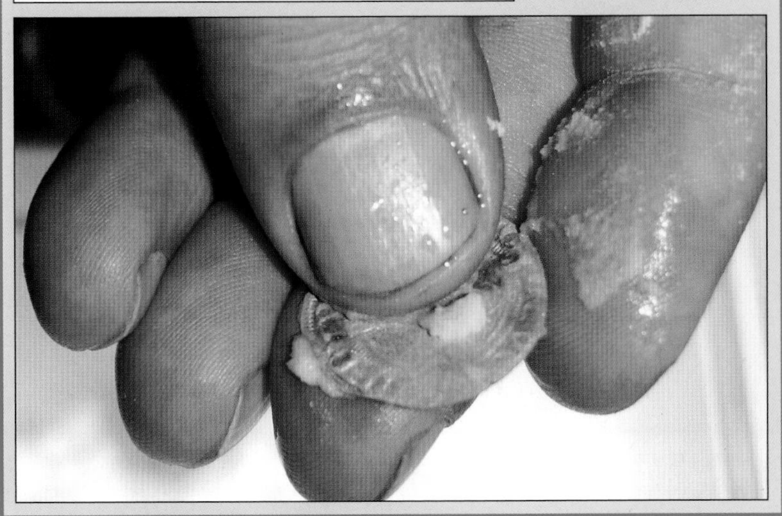

Leo first places a group of coins into this sugar water for about 12 hours. He suggests placing them into the water in the evening leaving them to soak overnight. The acid in the sugars eat away at some of the corrosion on the coins. In another small tub, he keeps a pile of the white sugary powder. He takes a pinch of this powder and then careful massages each coin with his fingers to gently remove more of the excess corrosion. Generally, some of the bad surface corrosion will begin to fall away.

After removing the coins from the sugar water and gently rubbing them with the powder, Leo then places them into a second

Leo removes a coin from the ammonia water bath to show the cleansing affect his process is giving to this found item.

(Right) Leo's final step to polishing up a rare coin is to gently apply an agent of his creation he calls "Patina Plus."

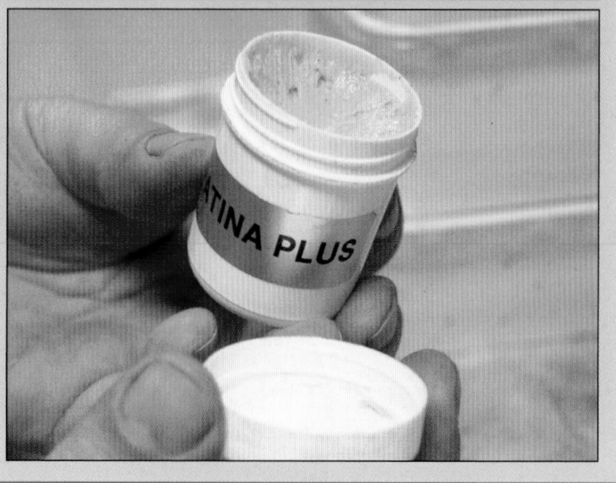

pan of distilled water (Step B). He leaves the coins in this more neutral solution for one hour.

Leo's third step is to place the coin next into another small tub that is filled with ammonia water (Step C). His ratio is 1 part ammonia to 3 parts water. He purchases a general ammonia used in household cleansers for this mixture. He then places his coins in this bluish-coloured water pan for another 12-hour period. After soaking overnight, he removes the coins from this mixture and again gently rubs them with the sugar/citrus powder.

To summarize:

1. Soak the coins for 12 hours in the citrus/sugar powder water and then gently rub away corrosion.

2. Move the coins into the distilled water solution for one hour.

3. Move the coins into the ammonia water solution for another 12 hours. Gently clean with your fingers and the citrus powder before rinsing them in distilled water.

4. Repeat the above three-step process a second or third time as needed to continue removing the corrosion.

5. The final step (Step D) when you are happy with the quality of your coins is to gently polish them by hand with a patina cleaner. Leo sells a special Patina-Plus gel to coin dealers and hobbyists which is ideal for cleaning bronze, silver, gold and copper coins. It is also good for cleaning any similarly composed metallic treasure to give it a final, clean shine. Your local coin dealers may sell a similar type solution. If you are unable to find such a gentle polishing gel, feel free to contact Leo through his website at www.kooistra-detectors.com/ to order his mixture.

Final "Polishing Up" Methods

As mentioned, techniques for cleaning old coins and artefacts vary widely, and great caution should be taken concerning the method you choose to employ. Some detectorists use common office supplies such as masking tape to remove tarnish from old coins. A high-tach masking tape can be used to remove some of the dirt and tiny debris from coin surfaces by carefully adhering and slowly peeling off clean pieces of tape from the coin. While this process will remove much of the dirt, be advised that it can also remove unstable patina.

Underwater treasure recovery expert Bob Marx offered another trick he uses to clean old Spanish reales. In his final cleaning process, he sometimes uses a simple rubber eraser to "erase away" the tarnish on big silver coins. Be aware, however, that the use of an eraser on newer coins can cause shadowing or scratches on the coin's face.

PHOTO QUALITY VARIANCE

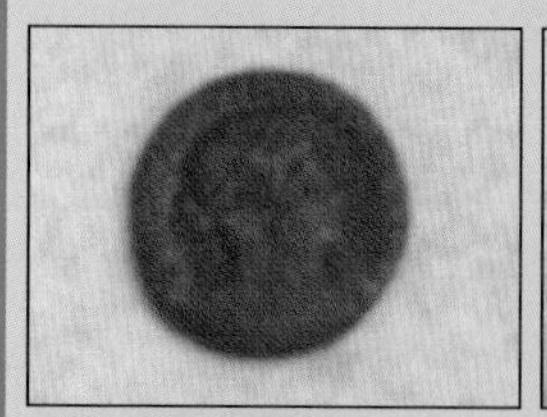

Each of these three photos were taken of the same coin. *(Left)* Photo taken with point-and-shoot digital camera shows lack of focus. *(Center)* Image taken with the same point-and-shoot camera under the same lighting using the camera's macro setting. *(Right)* Image taken with a higher quality digital SLR camera using a macro lens.

Photographing and Displaying your Finds

In some countries, the government may opt to keep your rare finds and offer some monetary compensation. Photographing your discoveries before reporting them is one way you can preserve the memory of a special item.

Digital single lens reflex (SLR) cameras with high resolution have become more affordable in the past few years. With a little experience behind the lens, such cameras enable you to take high quality photos with crisp detail. Basic non-SLR digital cameras can be more challenging for the operator to achieve good focus on a small object such as a coin.

When using a digital SLR or a point-and-shoot digital camera, always review the detail of your photo on the camera's LCD screen. If possible, zoom in on the photo to look closely at the image's sharpness. If the coin details appear shaky or fuzzy, delete the image and try again. Many digital cameras have a "macro" setting (often the symbol of a flower) that will help you capture the fine detail on such coins. Consult your camera's owner's manual about shooting extreme close-up photos. More serious photographers mount their camera on a tripod to keep it steady. Even the slightest shake as you push the button will throw the picture out of focus. Digital cameras can also capture the date when your photos are taken. This can be beneficial later if you maintain a journal or logbook of your finds.

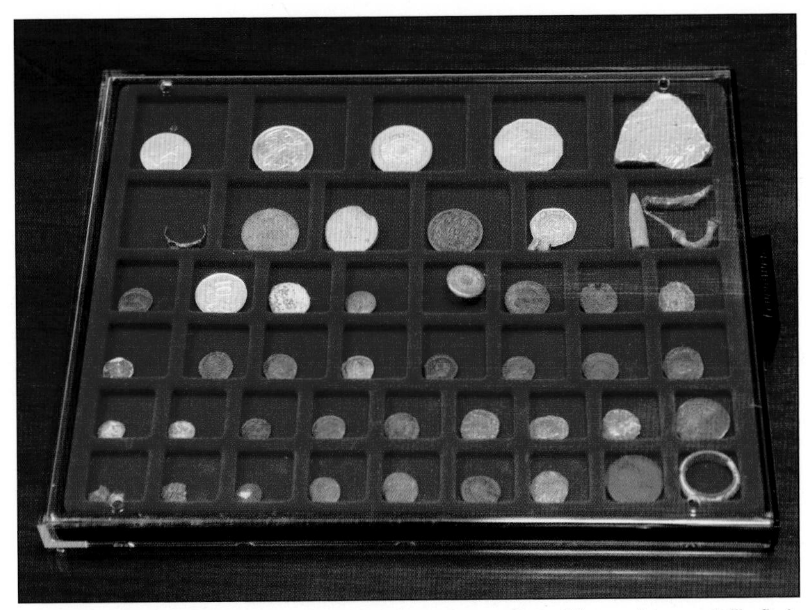

Commercially-available plastic coin and artefact trays such as this one keep metallic finds safely separated and protected from moisture.

Excessive handling of ancient coins can damage the surface appearance due to moisture and acids in your fingertips. Some detectorists opt to protect such valuable silver coins after cleaning by applying a thin coating of plastic lacquer for protection from handling. Coins can be stored in albums with plastic sheets. Be careful, however, while flipping pages that sheets of heavy coins do not damage the patina on small coins on adjacent pages.

If you opt to create your own display case, make certain to use acid-free materials and avoid allowing bare metal objects to come into contact with each other which can cause corrosion to spread. Items that are corroded should always be isolated from others. Another option is to use high quality plastic artefact and coin tray display cases. These are available from many metal detector dealers, coin dealers and even from some arts and crafts retailers. As you fill and collect more display trays you can further protect these trays from moisture by storing them in special cabinets built to house multiple trays or in airtight plastic storage tubs.

There are books available with complete details on restoration and preservation of coins and artefacts. Your best discoveries are the result of countless hours in the field. The extra effort you put into properly preserving them will enable you and your friends to enjoy these items for a lifetime.

CHAPTER 15

EUROPEAN TREASURE HUNTING LAWS

Treasure hunters new to the sport should certainly learn the laws that exist in their country or region of Europe before taking to the field with a metal detector. Many countries have adapted acts to promote the recording of chance finds and broaden public awareness of the importance of discovered objects.

The treasure laws pertaining to major European countries are listed below in an abbreviated format. Because such regulations are often amended, always check with the proper authorities in each particular country for any revisions.

Austria

Treasure hunting for archaeological items is generally off limits to private individuals. Such metal detecting is generally restricted by Austrian law to specific archaeological expeditions which require a special excavation permit issued by the Austrian Federal Monument Authority. The specific laws regarding general metal detecting for non-historic items should be checked before attempting a search in this country.

Belgium

The Belgian Code of Civil Law states that a landowner who finds a hoard on his own land has full rights to the treasure. If a hoard is found by someone on land other than their own, the finder must split the hoard equally with the landowner. Trespassing is forbidden, and any find made during an unauthorized entry onto another's land is completely forfeited by the treasure hunter to the landowner.

Coins, artefacts or other items of historical interest found during works on public properties become the property of the local authorities. The French-speaking Walloon Region of Belgium strictly prohibits metal detector usage by hobbyists, limiting the use of detectors to officially authorized excavations. Other regions of Belgium have passed decrees in recent decades pertaining to the archaeological use of metal detectors. It is therefore very important to look into the local decrees that exist within various regions before searching for treasure. In the Ardennes, a scene of heavy World War II conflict, it is illegal to search for military artifacts.

Bosnia

Treasure hunting is allowed but there is the obvious danger of the countless explosive remnants of war still in the ground throughout the country.

Bulgaria

Metal detecting is allowed in Bulgaria but unauthorized archaeological excavation and removal of historic artifacts in recent years has caused quite a political stir that may soon cause restrictions to be placed upon the use of detectors.

Cyprus

The Cyprus antiquities law of 1935, amended in 1973, states that no person can excavate on his land or anyone else's land for the "purpose of discovering antiquities without a license." The use of metal detectors is not specifically mentioned but the strictness of this law rules them out except where a person has been licensed to dig for antiquities.

Czech Republic

Metal detecting is legal on private and public properties in the

Czech Republic with the exception of national monuments and archaeological sites. Historic artefacts that are discovered must be reported according to Czech law. The finder will be paid 100 per-cent of the value of items of precious metals, and 10 per cent will be paid for other materials, such as stone, wood, and ceramic. In addition, Czech detectorists who report on an unknown archaeo-logical site will be paid a finder's fee.

Denmark

Metal detecting is allowed in Denmark with a number of restrictions that must be noted. Historical and archaeological sites are strictly forbidden areas for metal detectors. Local communities have the right to determine if public land can be hunted with metal detectors; approximately half of all public lands are closed.

Private lands (with the owner's permission) and public beach areas are open to metal detecting. Any coins found that were mint-ed after Denmark's coin reform in the 19th Century may be kept. All earlier coins or artefacts must be delivered to the National Mu-seum, which awards a cash sum for the find. Treasure hunters are rarely allowed to keep their older finds and the awards offered for older coins are generally below market value. An old law from 1421 called "Danefae" which states that says that gold or silver anywhere belongs to the king is still enforced.

In a newer 2001 Museum Act of Denmark, Section 9 defines items considered to be treasure trove, objects of the past which no one can prove to be the rightful owner. "Treasure trove shall belong to the state. Any person who finds treasure trove, and any person who gains possession of treasure trove, shall immediately deliver it to the National Museum of Denmark. The National Mu-seum shall pay a reward to the finder. The amount shall be fixed by the National Museum on the basis on the value of the material and rarity of the find and also of the care with which the finder has safeguarded the find." Persons failing to report such discoveries will be fined.

England and Wales

The British Treasure Act of 1996 was passed to replace the medieval law of Treasure Trove in England and Wales which had given some protection to certain archaeological finds. The Treasure Act led to the a visionary reporting plan known as the *Portable Antiquities Scheme,* created to record archaeological objects found by members of the public in England and Wales. Six pilot schemes were established and based in museums and archaeology services in Kent, Norfolk, the West Midlands, North Lincolnshire, the North West and Yorkshire. These six posts and schemes were funded and coordinated by the British Museum in London. A Finds Liaison Officer (FLO) was placed in each of the six pilot areas to work with detectorists and their discoveries.

More than 13,500 historical objects were found during the first year of the pilot scheme, and this led to extensions of the program to provide a comprehensive national plan for all of England and Wales. At present, there are 36 FLOs who cover each county in England and all of Wales, plus additional advisers to work with detectorists.

The British definition of treasure, according to the Treasure Act, outlines what must be reported. These guidelines were established in 1997 and amended in 2003 to include prehistoric artifacts.

What is considered treasure that must be reported?

• Any metallic object, other than a coin, that is at least 300 years old and contains *at least 10 per cent* by weight of a precious metal (gold, silver).

• Any single prehistoric item will be considered treasure, provided that *any* part of it contains a precious metal.

• Any group of *two or more* metallic items of any composition that are of prehistoric date and that are *found together*.

• Groups of coins coming from the same find that are at least 300 years old. If the coins contain less than 10 per cent gold or silver, then there must be at least ten coins found together to be considered a treasure. **Coins coming from the same find are defined as:** hoards that were deliberately hidden; smaller groups of coins dropped or lost together (such as a coin purse contents); and votive or ritual deposits.

• Other objects, regardless of precious metal content, that are found together with an object defined as treasure (see above rules) or an object that had previously been together with the treasure item. Because finds may have scattered since they were originally deposited in the ground, objects found that appear to have at one time been part of the treasure would be grouped together as treasure.

• Items less than 300 years old made substantially of gold or silver that were deliberately cached with their owner now unknown. This last article covers items that would have originally been a treasure trove object but are not covered by the other rules stated above.

Metal detectorists who find any treasure item that falls within the above described parameters must report their discovery to a district coroner within 14 days of making the discovery—or within 14 days after the find is determined to be treasure. The penalty for failing to report a treasure discovered in England or in Wales is a criminal offence whose maximum punishment can be up to three months in jail or a fine not to exceed level 5 (current £5,000), or both.

Estonia

This tiny country broke away from the Soviet Union in 1992 and became a member of the European Union (EU) in 2004. The northeast of Estonia is rich in World War II artefacts where WWII "relic hunting" has become popular. As for detecting coins and artefacts dated pre-19th century, the area around Tallinn (Reval), the capital of Estonia and the areas along the southern border with Latvia are known to be productive. As in all European countries, current laws as they regard to antiquities should always be consulted.

An avid treasure hunter who uses the Internet pseudonym "Sergei from Upstate New York" is a Russian native who enjoys

When Sergei walked with his mother along the Baltic Sea shore, he came across this abandoned homestead.

Photos courtesy of Sergei and his website:

www.metaldetectingworld.com

Coins from two ruling countries such as Sweden and Russia can be found in Estonia.
(Left) Sergei recovered some of these coins at the old homestead site seen above. The coins range in date from a 1906 silver 20-Kopek to the smaller 1-Kopek silver hammered coins at bottom, which were minted from the 1480s to the 15402.

traveling back to Europe to metal detect with his friends in Russia, the Ukraine and in Estonia. He points out that searching in Estonia often produces coin finds from the neighboring countries of eastern and northern Europe. "The area that includes Estonia, Latvia and Lithuania was historically under one king or queen or one name, such as Livonia, in medieval times," Sergei noted. "I have found many Russian coins, circa the 15th–19th centuries, in Estonia but just recently I found in the neighboring region of Russia a Swedish hammered silver 1649 1-Öre [the upper right silver coin in the above photo] that was minted in Reval during Swedish occupation."

Finland

Metal detecting is not ruled out, although the country's 1963 Antiquities Act declares that all moveable objects—such as coins, artefacts or weapons—more than 100 years old must be reported.

France

A new law was passed in December 1989 to prevent the public from searching national monuments for artifacts of historical interest without obtaining a proper license. Even general metal detecting in France requires users to obtain such a permit before hunting. When searches are to be conducted on land not belonging to the applicant, the written application must be accompanied by a document of consent written by the landowner.

To obtain such a permit, the person must complete an authorization application that includes such information as the identity, competence and experience of the applicant. The location, scientific objective and the duration of the search must also be detailed and written consent from the landowner(s) must accompany the application. Anyone caught metal detecting without the proper permit is subject to being fined and to having their equipment confiscated.

This law is not believed to prevent hobbyist from using metal detectors on public beaches—with the obvious exception of beaches of historic interest such as those of Normandy. More stringent laws were passed in 1999 to prevent the exportation of cultural property.

Germany

Metal detecting is permitted in the 13 federal-states and three city-states, with certain restrictions regarding the discovery of antiquities. A 1992 law on the search for, and preservation of, antiquities states that a license is required to search for objects belonging to the ancient period, early Christianity and the Middle Ages. Accidental discoveries must be reported. The government offers a reward equal to 50 percent of value for items found on public lands and 100 percent of value if the item was found on private land. It is forbidden to look for buried archaeological items in all German states and this includes World War II artefacts in some areas of the country.

Greece

Greece's 1932 antiquities law made all artefacts on land and in the sea belong to the state, but it has not outlawed treasure hunting off its 1,500 kilometers (9,400 miles) of coastline. About 100 known underwater archaeological sites are closed to treasure divers and Greece works to pay handsome rewards for other artefacts that are discovered by fishermen or divers. In general, metal detecting on land is banned in Greece.

A law of 2002 further defines the protection of Greek antiquities and cultural heritage. Article 8 states that anyone who finds an antiquity must declare it "to the nearest archaeological, police or port authority." Rewards may be paid based upon the significance of the discovery. Article 38 states that "the use of metal detectors or other scanners for surveying the subsoil, seabed or bed shall not be permitted without a permit." A person using a metal detector without such authorization, according to Article 62, "shall be punished by a term of imprisonment of not less than three (3) months." Habitual violators who are caught shall be imprisoned for no less than three years.

Hungary

This is a very popular country for metal detecting. Strict laws are in place, however, due to the significant number of archaeological sites in this country.

Ireland

The Treasure Act of 1996 replaced the old common law of treasure trove that had been in place in England, Wales and Ireland. In the Republic of Ireland, a 1987 amendment to the National Monuments Act prohibits a person from metal detecting at the site of a national monument or a registered archaeological area. It is also illegal in Ireland to advertise or promote a metal detector for the purpose of searching for archaeological objects.

In Northern Ireland, metal detecting is prohibited unless the individual has obtained a license from the Department of the En-

European archaeological sites are considered national monuments in which metal detecting is forbidden. Each country has specific laws concerning chance finds of historic artifacts.

vironment, with the stipulation that all finds be reported. Anyone finding an archaeological item must report it within 14 days to the Director of the Ulster Museum or to the officer in charge of a police station.

Italy

Metal detecting is forbidden in certain areas of Italy: Val D'aosta, Toscana, Lazio, Calabria and Sicilia. In the areas of Italy where metal detecting is allowed, all objects and coins of historic or archaeological interest must be reported within 36 hours to the Superintendency of Arts and/or local authorities. These objects become state property, although rewards may be offered up to one-quarter of the artefact's value.

Coin hunting is popular in the Netherlands, where old Dutch windmills dot the northern countryside.

Coin hunters can keep coins minted after 1500 but 10 percent of the coin's value or 10 percent of your finds must be paid to the landowner. The land owner must give you permission *each time* you hunt his or her property. Beachcombing is generally allowed as long as detectorists follow the general rules of reporting significant finds.

Latvia

Specific laws are unknown although Latvia has legislation for severe penalties—up to five years confinement or a fine of up to one hundred times the minimum monthly wage—for persons who destroy or damage sites under special State protection.

Liechtenstein

A governmental permit is required for any archaeological excavations. The country's 1977 Monument Protection Act states that any antiquities found in the ground must be declared to proper government authorities.

Lithuania

Metal detecting is allowed in Lithuania, where a new state system of cultural heritage management has been active since 1995. As of 2001, approximately 19,000 objects, complexes and sites of cultural heritage value have been registered with the Department of Cultural Heritage Protection. As in all European countries, the current antiquities laws should always be consulted.

Luxembourg

According to a 1966 act on excavations and moveable cultural objects, any search or excavation that hopes to unearth items of historical interest can be carried out only with the authorization of the Minister for Arts and Sciences.

Malta

The 1925 Antiquities Protection Act protects all objects, both movable and immovable, which are more than 50 years old. This act has undergone numerous revisions, the most recent being in 2001. Accidental finds must be reported and excavation can only be carried out with government authorization. Metal detectors possessing sufficient sensitivity to be of danger to archaeological sites have been banned in Malta since 1979.

The Netherlands (Holland)

Metal detecting is big in the Netherlands. This area, once ruled by the Roman Empire, is rich with history and offers much for the detectorist. According to the country's antiquities laws, detectorists must report to authorities and turn in coins and objects of historical interest that are more than 50 years of age. Local hobbyists report, however, that the government is primarily interested in objects that predate 1700. Two Dutch regions, Nijmegen and Arnhem, have placed a ban on metal detecting. A 1984 Cultural Heritage Protection Act protects items of historical interest from being exported.

Norway

Laws in this country do not specifically reference metal detectors but speak generally to the handling of historic items. A wide variety of items dating from before 1937, both fixed and moveable, are protected by the Cultural Heritage Act of 1978, which also provides protection from unauthorized excavation. The ownership of all objects older than 1537 and of coins older that 1650 is vested in the State (Section 12, a and b). Section 13 requires that all finds should be reported to the authorities who will fix a suitable reward. There is no specific reference to metal detectors. Cultural objects are protected from export by a regulation of January 2007.

Poland

Metal detectors for hobbyists became available here during the 1980s. Since 2003, detectorists must apply for permission from their local authorities, and items of archaeological importance must be reported. As in other European countries, the government offers financial rewards for items that are kept.

Portugal

Metal detecting is not allowed inland because of the significant number of archaeological sites. The country's beaches are generally the best place to search because you encounter the least restrictions here.

Romania

Metal detecting is against the law in Romania without a special permit. Legislation passed since 2001 imposes heavy monetary penalties for even owning or selling detectors in Romania without a required permit. Only registered archaeologists can conduct work on known sites of historical interest. A "chance archaeological find" where heritage items are uncovered must be reported by the owner of the land to his or her town mayor within 72 hours.

The use of metal detectors in archaeological sites "is allowed only on the basis of an authorization already issued by the Ministry of Culture and Religious Affairs." Violation of this law is punishable by confiscation of the detectors and one to five years in prison. Romania has also established a National Archaeology Commission which has no legal status but serves in a consultancy role in the field of archaeological heritage.

The rolling hills in Spain are often rich with history but the country's antiquities laws are very strict concerning the use of metal detectors.

Scotland

The Scottish Crown claims all archaeological finds made in this country, whether they are composed of precious metals or not. All objects found must be reported to the person's local museum for examination. In some cases where the object is not desired for a museum collection, it is returned to the finder.

Slovakia

Metal detecting is allowed; specific laws are unknown.

Spain

The use of metal detectors in public areas requires a permit from local authorities. Historical finds, particularly those of an item more than 100 years old, must be reported to authorities.

Spain has become more stringent concerning metal detector use since 2001, and a number of detector users have faced prosecution for removal of historical objects. Your best bet in some areas is to participate in rallies whose organizers have obtained special permission to hunt—under the stipulation that the recovered items are handed over to authorities.

Sweden

Metal detecting was banned in the counties of Gotland and Oland in 1988, and this ban has since been extended to include all of Sweden. Metal detecting within Sweden's 25 states is allowed only for individuals who have been given special permission to do so for a specified time. All finds made by such individuals must be reported to the National Heritage Board by county administrative boards, county museums, the Coast Guard or police authorities.

Switzerland

The unauthorized search or excavation of antiquities has been banned by Swiss legislation, but general metal detecting by private individuals has not been significantly harmed by legislation. A federal act of 2003 prohibits the exportation or distribution of cultural items outside of Switzerland; fines of up to 100,000 Swiss francs or one year in jail can be imposed.

Turkey

Metal detecting laws are very strict in this country and metal detecting appears to be forbidden unless a special license has been granted by the government.

Chapter 16

Metal Detectors and Archaeology

Even though metal detectorists turn up history, they have not always been looked upon favorably by professional archaeologists. Still, progressive countries have taken steps to help open communication between amateur "relic hunters" and archaeologists. Britain is among these nations that have encouraged treasure seekers to turn in their finds to proper authorities. Since the 1996 Treasure Act, detectorists have been offered market value for their discoveries. The British Museum has the first option to buy the treasure or can opt to return it to the finder.

The willingness of detectorists to report finds and claim their rewards has shown promising growth. In 1999, there were only about 25 "treasure" recoveries reported to British officials. By 2008, however, the number of finds reported had grown to 802. Roger Bland of the British Museum in London admitted in 2009, "The collections in our museums would be thinner without the detectorists' finds." British archaeologists and county recording officers now offer talks to some of the regional detecting clubs aimed at improving the detectorists' understanding of their work.

In many other European countries, this system is not as productive. Detectorists, in fact, often feel so threatened by the legal system that they fear reporting a treasure find. In some cases (as seen in the previous chapter), only a nominal fee is paid for a reported treasure...or perhaps no fee at all. Thus, some finders generally opt not to turn in their recoveries.

Today's archaeologists often use metal detectors as part of their process in uncovering the remnants of an historic area. In many cases, the professionals turn to experienced metal detectorists to help them in their searches. Mike and Gail Bartley have long enjoyed a working relationship with historians and archaeologists thanks to their work with the Northumbrian Search Society, a British club founded in 1975 to help preserve history.

"Our group of detectorists works closely with the archaeologists on exciting projects," says Gail. "In fact, we often get to go to sites that are off limits for metal detecting." The Northumbrian Search Society's detectorists have helped recover coins, buckles and various Victorian-era artefacts while working on their projects. Gail is pleased with the positive steps that have been made in recent years between the professional and amateur historians. "There is a new breed of archaeologist out there now willing to work cooperatively with detectorists."

Archaeological discoveries create a certain buzz of excitement in both the historical and amateur metal detecting circles. This was certainly evident during the September 2009 West Hanney rally near Oxford when a Saxon grave was located. Chris Bayston from Yorkshire pursued a deep signal that proved to be an ornate brooch that was likely buried with an early member of a Saxon royal family.

"Archaeologists say that the person in the grave at the West Hanney rally would certainly have been one of high status and probably of royal lineage," relates Weekend Wanderers rally organizer Pete Welch. "The most notable Saxon King in the area was King Alfred, who was born at Wantage just 4 miles away. It is a possibility that the person buried at West Hanney is a direct descendant but at this moment there is no way of proving that."

According to Welch, the 4-inch wide brooch is "one of only 20 known brooches of this type found in England. This one is the most western discovery so far. It was made in southeast England in the county of Kent and is therefore known as a Kentish composite brooch. The garnets, which are quite comparatively large, date

(Above) Archaeologists work to uncover the remains of a Saxon grave site near West Hanney, England, in September 2009. The skeletal remains are believed to be those of someone of high status, possibly royalty. To one side of this seventh century skeleton are two earthenware pots which were placed atop two iron knives *(see below left)*.

The West Hanney remains were buried with a rare brooch. " A spindle whorl and two pieces of blue/green glass were placed between the legs," noted Pete Welch, organizer of the rally in which the discovery was made. "Unusually, there was not a second brooch; they are often found in pairs. Early inspection of this skeleton while in situ shows that the bones were fused, proving that the person was at least a young adult and likely female."

Photos courtesy of Pete Welch of Weekend Wanderers Detecting Club of England.

the brooch to the early seventh century. Garnets in later brooches were smaller as the stones became more expensive. The garnets are underlaid with gold to enhance luster. The white material in each boss is shell that came from the Red Sea and is set in an intricate gold wire support which is then surrounded by red garnets. Between the turrets are gold plaques inlaid with scroll work designs. Soil removed from the grave will be washed and hopefully loose stones and gold from the brooch will be recovered and the brooch can be restored to its full glory."

Archaeologists carefully inspected the soil around the skeleton for other artefacts. They lifted to body to scan below it and then reinterred the remains. Ground penetrating radar will be used in the area to assess if there are other Saxon graves. The excavation and assessment work of such an important find is tedious but the results will certainly add to the historical knowledge of early southeastern England history.

How Can You Get Involved with History?

• *Volunteer as a steward or detectorist with an archaeological project.* In most cases, there is no pay involved and plenty of hard work in the field. The reward is in preserving history and witnessing the items from ancient civilization that are unearthed.

• *Join a metal detector club that is involved with historic projects where you can lend your services.* Archaeologists often train interested people in the proper techniques of preserving artefacts recovered in the field.

• *Conduct your own research to pinpoint areas of special historic interest, making certain that you are not interfering or trespassing on areas that have been previously classified as archaeological sites.* Any area where people once worked or lived can yield coins and other interesting finds. Follow the laws that apply to your country if you discover significant cultural artefacts. Your discoveries could prove to be of great interest to archaeologists.

Italian archaeologists using a Garrett *Sea Hunter* metal detector off the island of Capraia made a discovery that surprised them. This ancient pottery, dating back 2,150 years, is actually detected by their *Sea Hunter* because of the iron minerals in the clay that was used to form the pottery thousands of years ago. Marine archaeologist Dante Bartoli believes the pottery may have been produced at Campania, the volcanic area of Italy near Vesuvius.

In Italy, a group of volunteer detectorists are being organized by Aldino Bartolini to work with police forces and archaeologists to conduct hunts when they are needed. These volunteers must be proficient with their metal detection equipment and are prescreened by authorities. Once they have been certified, the volunteer detectorists are then issued a photo ID that allows them to work with archaeologists and other authorities. "It is worthwhile because we can now hunt for artefacts in places where we would not otherwise be allowed to do so," Bartolini says.

The Archaeology Process

The process in which metal detectors are incorporated into archaeological excavations varies according to how finite the search area has been specified. For larger tracts of land where historians hope to pinpoint a specific settlement, battle engagement or other significant happening, they must first narrow down the area.

Volunteer or government-authorized detectorists generally begin by conducting a reconnaissance-level survey of the project area. Such "recons" are not expected to provide 100% coverage of large areas; rather, they hope to use a number of detectorists to range over areas of interest until hot spots are encountered. In areas where artefact concentrations are discovered during these initial surveys, more refined block or radial survey zones are then established to work over these areas of concentration more thoroughly.

Metal detector operators are expected to be proficient with their equipment, particularly able to properly ground balance their machines to the soil content that is present. Such calibrations might range from normal soil to areas of wet sand, high salt soil or highly mineralized ground. Some archeological guidelines further specify that searchcoils should be at least 23 cm (9-inches) in diameter and that headphones are required for each user. The detector's sensitivity should be raised to the highest settings allowable based on soil conditions, and the detector's volume should also be set to the highest comfortable range possible. In areas of artefact concentration, it is especially important that detectorists overlap searchcoil sweeps by 50 percent to avoid skipping any good targets.

In preparation for reconnaissance-level surveys or intensive grid zone searches, the project site should be cleared of tall weeds and shrubs. It is important that searchcoils are able to sweep as close to the earth as possible in order to achieve satisfactory ground penetration. Search grids are often marked with strings tied off to corner stakes, or the ground is painted with a suitable florescent-colored paint. The detectorists are generally urged to maintain at least a three-meter distance from each other to avoid interference between machines.

All metal discoveries, or "hits," are dug for identification. The recovery team usually consists of a volunteer digger, or person who handles the excavation, and an archaeological steward, the person who records data associated with each find and tags the item. The artefacts are dug carefully to avoid damage from the shovel or hand tools. The digger will start an excavation several centimeters behind and around the target area that the detectorist has pinpointed. The excavated soil and any vegetation is carefully replaced after the item or items have been removed.

A field form is completed as artefacts are removed. Each item is given a unique location number to help map its coordinate point in relation to the block or zone it was recovered in. The steward also carefully records the depth of the item, the name of the person who pinpointed it, the date and other relevant details. The item is placed in a labeled bag, and the excavation location is mapped by GPS, conventional survey or some other approved method.

The recovered artefacts are delivered to the lab that is handling the conservation for that particular project. The items will often first receive a simple rinsing under fresh water and possibly a light scrubbing with a soft brush to remove as much of the soil as possible. The items are then dried and re-bagged for more extensive cleaning as required.

The final conservation of the specimens is handled under the supervision of a specialist in artefact conservation. The best way to experience the process of a true archaeological project is to volunteer your services to regional experts. The labor of your searches might just prove to be something from which your grandchildren will benefit.

DETECTING WITH ARCHAEOLOGISTS: THE PROCESS

(Above) A team of Garrett metal detectorists participated with professional archaeologists on a Texas Revolution battleground that is now a State Historic Site. Archaeologist Roger Moore shoots in an artefact from the total station (base).

(Below) The field team lines up the artefact by placing their site pole in the recovery hole as detectorist Herman Denzler takes a break from the midday heat.

Note: Garrett metal detectors were used at San Jacinto Battleground State Historic Site in Texas during controlled archaeological investigations and under the supervision of professional archaeologists. Recreational use of metal detectors is prohibited by law at all Texas State Parks and Wildlife Management Areas.

Steve Moore enjoys the recovery of a musket ball *(right and below)* on the 1836 battleground where two of his ancestors helped Texas win its independence.

The artefact recovery process. *(Left)* Detectorists watch as the stewards dig and pinpoint artefacts.

The volunteer steward *(below, left)* records the data on a form for each artefact and then seals them in a bag *(below)* that is also marked with the data.

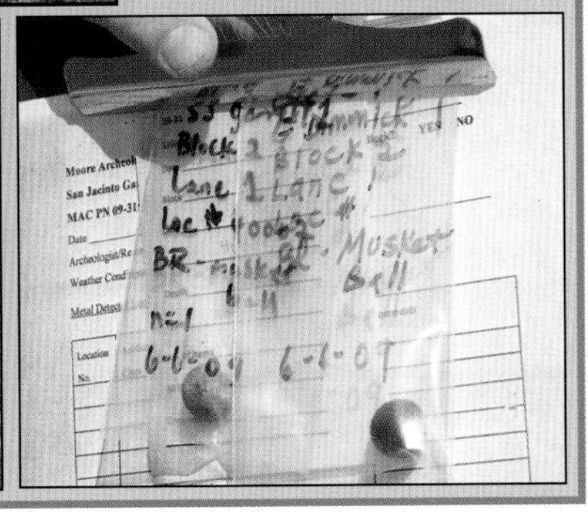

(Right) Charles Garrett shows his enthusiasm as his team recovers another artefact. *(Below)* This simplified image from the archaeologist's software shows how they learn from the plotted concentrations of artefacts recovered at San Jacinto.

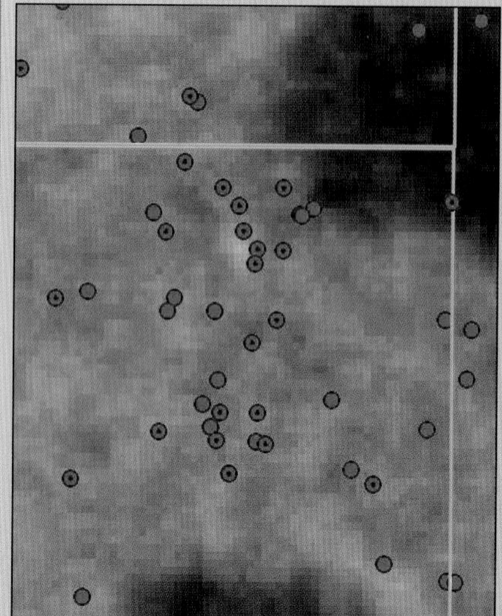

By participating in professional archaeological projects, the detectorists have the satisfaction of helping to recover significant history that can soon be enjoyed by others in museums.

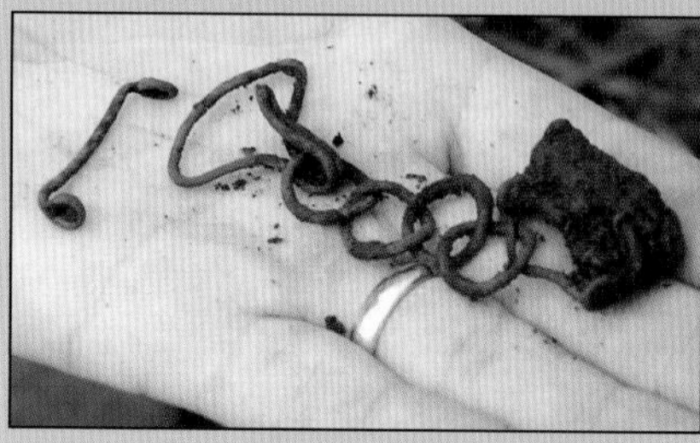

Among the many items Garrett detectorists helped unearth on this project were *(clockwise from above left)*: a chain from an officer's uniform, a cluster of dropped musket balls found by the author, a buckle from a shot pouch, and this canister base from an artillery shell.

CHAPTER 17

METAL DETECTORIST'S CODE OF ETHICS

Today's dedicated metal detector hobbyist must do more than simply fill in excavation holes and protect landscaping. The antiquities laws are tough in many European countries and will only become tougher if individuals and organizations do not follow a strict code of ethics.

Several codes of conduct have been written by metal detector groups in Europe and in the United States over the years. More recently a voluntary "Code of Practice for Responsible Metal Detecting in England and Wales" has been adopted by a number of organizations. They include: the British Museum; the Council for British Archaeology; Country Landowners and Business Association; English Heritage; Federation of Independent Detectorists; Museums, Libraries and Archives Council, National Council for Metal Detecting; National Farmers Union; National Museum Wales; Royal Commission on the Historic and Ancient Monuments of Wales; Portable Antiquities Scheme; and the Society of Museum Archaeologists. This code states:

CODE OF PRACTICE FOR RESPONSIBLE
METAL-DETECTING IN ENGLAND & WALES

Being responsible means:

Before you go metal-detecting:

1. Not trespassing; before you start detecting, obtain permis-

sion to search from the landowner/occupier, regardless of the status, or perceived status, of the land. Remember that all land has an owner. To avoid subsequent disputes it is always advisable to first get permission and agreement in writing regarding the ownership of any finds subsequently discovered (see www.cla.org.uk or www.nfuonline.com).

2. Adhering to the laws concerning protected sites (e.g. those defined as Scheduled Monuments or Sites of Special Scientific Interest: you can obtain details of these from the landowner/occupier, Finds Liaison Officer, Historic Environment Record or at www.magic.gov.uk). Take extra care when detecting near protected sites: for example, it is not always clear where their boundaries.

3. You are strongly recommended to join a metal detecting club or association that encourages co-operation and responsive exchanges with other responsible heritage groups. Details of metal detecting organisations can be found at www.ncmd.co.uk or www.fid.newbury.net.

4. Familiarising yourself with and following current conservation advice on the handling, care and storage of archaeological objects (see www.finds.org.uk).

While you are metal-detecting:

5. Wherever possible working on ground that has already been disturbed (such as ploughed land or that which has formerly been ploughed), and only within the depth of ploughing. If detecting takes place on undisturbed pasture, be careful to ensure that no damage is done to the archaeological value of the land, including earthworks.

6. Minimising any ground disturbance through the use of suitable tools and by reinstating any excavated material as neatly as possible. Endeavour not to damage stratified archaeological deposits.

7. Recording findspots as accurately as possible for all finds (i.e. to at least a 100m2, using an Ordnance Survey map or hand-held Global Positioning Systems (GPS) device) whilst in the field. Bag

finds individually and record the National Grid Reference (NGR) on the bag. Findspot information should not be passed on to other parties without the agreement of the landowner/occupier (see also clause 9).

8. Respecting the Country Code (leave gates and property as you find them and do not damage crops, frighten animals or disturb ground nesting birds, and dispose properly of litter, see: www.countrysideaccess.gov.uk)

After you have been metal-detecting:

9. Reporting any finds to the relevant landowner/occupier; and (with the agreement of the landowner/occupier) to the Portable Antiquities Scheme, so the information can pass into the local Historic Environment Record. Both the Country Land and Business Association (www.cla.org.uk) and the National Farmers Union (www.nfuonline.com) support the reporting of finds. Details of your local Finds Liaison Officer can be found at www.finds.org.uk, email info@finds.org.uk or telephone +44 (0)20 7323 8611.

10. Abiding by the provisions of the Treasure Act and Treasure Act Code of Practice (www.finds.org.uk), wreck law (www.mcga.gov.uk) and export licensing (www.mla.gov.uk). If you need advice, your local Finds Liaison Officer will be able to help.

11. Seeking expert help if you discover something large below the ploughsoil, or a concentration of finds or unusual material, or wreck remains, and ensuring that the landowner/occupier's permission is obtained to do so. Your local Finds Liaison Officer may be able to help, or you should seek the advise of an appropriate person. Reporting the find does not change your rights of discovery, but will result in far more archaeological evidence being discovered.

12. Calling the Police, and notifying the landowner/occupier, if you find any traces of human remains.

13. Calling the Police or HM Coastguard, and notifying the landowner/occupier, if you find anything that may be a live explosive: do not use a metal-detector or mobile phone nearby as this

might trigger an explosion. Do not attempt to move or interfere with any such explosives.

Metal detector organizations and local clubs across the world have adopted similar policies for proper conduct. Charles Garrett has for many years published a basic code of ethics in his treasure hunting books:

- I will respect private and public property, all historical and archaeological sites and will do no metal detecting on these lands without proper permission.
- I will keep informed on and obey all laws, regulations and rules governing federal, state and local public lands.
- I will aid law enforcement officials whenever possible.
- I will cause no willful damage to property of any kind, including fences, signs and buildings, and will always fill holes I dig.
- I will not destroy property, buildings or the remains of ghost towns and other deserted structures.
- I will not leave litter or uncovered items lying around. I will carry all trash and dug targets with me when I leave each search area.
- I will observe the Golden Rule, using good outdoor manners and conducting myself at all times in a manner that will add to the stature and public image of all people engaged in the field of metal detection."

Policing these codes is an important job of the local metal detector and treasure hunting clubs organized throughout Europe.

SELECTED BIBLIOGRAPHY

ARTICLES, JOURNALS AND PAMPHLETS

"Brits Really Dig Generous Finder's Fee." *The Dallas Morning News*, May 15, 2009, 16A.

Bailey, Bob. *Metal Detecting Farm Holiday Guide.* Halifax: Prime Print, undated pamphlet.

Jack, Malcolm. "Top 10 Metal Detector Discoveries." *Heritage Key*, September 24, 2009.

Osborne, Chris. "Two Greenhorns try Panning for Gold." *The Searcher*, September 2009, 20–21.

Teplyakov, Sergei. "From Russia...With Relics. An American Goes Abroad to Recover the Distant Past." *American Digger*, July–August 2007, 27–30.

"The Vale of York Hoard." *The Searcher*, November 2009, 9.

BOOKS

Appels, Andrew and Stuart Laycock. *Roman Buckles & Military Findings.* Witham, Essex: Greenlight Publishing, 2007.

Bailey, Gordon. *Buttons & Fasteners: 500 BC—AD 1840.* Witham, Essex: Greenlight Publishing, 2004.

——. *Detector Finds 3.* Witham, Essex: Greenlight Publishing, 2001.

Cherry, John. *The Middleham Jewel and Ring.* Museum Gardens, York, UK: The Yorkshire Museum, 1994.

Evan-Hart, Julian and Dave Stuckey. *Beginner's Guide to Metal Detecting*. Witham, Essex: Greenlight Publishing, 2007.

Garrett, Charles. *The New Successful Coin Hunting.* Garland, Tex.: RAM Books, 2005 (Thirteenth Revised Edition).

——. *Treasure Hunting for Fun and Profit.* Garland, Tex.: RAM Books, 2005 (Eighth Printing).

——. *Understanding Treasure Signs and Symbols.* Garland, Tex.: RAM Books, 2009.

——. *Introduction to Metal Detecting in Europe.* Garland, Tex.: RAM Books, 2009.

Gesink, Gert. *Handbook for Detectorists.* The Netherlands: Detect, Enschede, 2005.

Grun, Bernard. *The Timetables of History. The New Third Revised Edition.* New York: Simon & Schuster, 1991.

Howland, John. *Treasure From British Waters.* Garland, Tex.: RAM Books, 1991.

Langer, William L. (editor). *An Encyclopedia of World History.* Boston: Houghton Mifflin Company, 1960.

Marx, Jenifer. *Gold in the Ancient World: How it Influenced Civilization.* Garland, Tex: RAM Books, 2009.

Marx, Robert F. and Jenifer. *The World's Richest Wrecks: A Wreck Diver's Guide to Gold and Silver Treasures of the Sea.* Garland, Tex: RAM Books, 2009.

Mills, Nigel. *Celtic & Roman Artefacts.* Witham, Essex: Greenlight Publishing, 2007.

Sear, David R. *Roman Coins and Their Values, Vol. I.* London: Spink & Son, Ltd., 2000, 5th edition.

——. *Roman Coins and Their Values, Vol. III.* London: Spink & Son, Ltd., 2005.

Skingley, Philip (editor). *Coins of England & the United Kingdom.* London: Spink & Son, Ltd., 2009, 44th edition.

Villanueva, David. *Cleaning Coins & Artefacts.* Witham, Essex: Greenlight Publishing, 2008.

Willemsen, Annemarieke. *Vikings! Raids in the Rhine/Meuse Region, .800–1000.* Utrecht: Centraal Museum, 2004.

Wolf, Norfolk. *Advanced Detecting.* Witham, Essex: Greenlight Publishing, 2005.

INTERNET SOURCES

"The Staffordshire Hoard. The Largest hoard of Anglo-Saxon gold ever found." Accessed http://www.staffordshirehoard.org.uk/artefacts/ on September 24, 2009.

Wessex Metal Detecting Association of Newbury, Berkshire (UK). Accessed http://wessex.club.googlepages.com/home on July 10, 2009.

Web Resources on Cultural Property, Antiquities Laws and Related Subjects. Accessed http://www.ulb.ac.be/assoc/aip/webresources.htm on July 23, 2009.

Bruce, Sarah. "On first time out with his metal detector, amateur treasure hunter finds £1m hoard of ancient golden jewellery." Accessed http://www.dailymail.co.uk/news/article-1225271/David-Booths-1m-gold-Stirlingshire-Amateur-treasure-hunter-finds-hoard-ancient-jewellery.html on November 9, 2009.

ABOUT THE AUTHOR

Stephen L. Moore, a sixth generation Texan, is the author of ten previous books on World War II and Texas history. His Texas history titles include *Savage Frontier*, a multi-volume chronology of the early Texas Rangers, and *Eighteen Minutes: The Battle of San Jacinto and the Texas Independence Campaign*. Moore, a frequent speaker at book conferences and metal detecting club events, writes for local historical journals, including *The Texas Ranger Dispatch*.

In recent years, he has combined his passion for history with the use of metal detectors to promote responsible relic recovery. His most recent title from RAM Books, *Last Stand of the Texas Cherokees*, details how he and a team of detectorists were able to pinpoint key locations of an important 1839 battle. Steve currently serves as the marketing and advertising manager for Garrett Metal Detectors. He, his wife Cindy and their three children live north of Dallas in Lantana, Texas.

THE GARRETT LIBRARY

These standard-size 5.5" x 8.5" format books offer treasure hunting techniques, hints and history from Charles Garrett and other RAM Books authors. Each book is soft cover format unless otherwise noted.

Visit garrett.com often to watch for new titles!

PN 1501500

PN 1508600

PN 1508200

PN 1508300

PN 1501700

PN 1545500

PN 1505470

PN 1500000

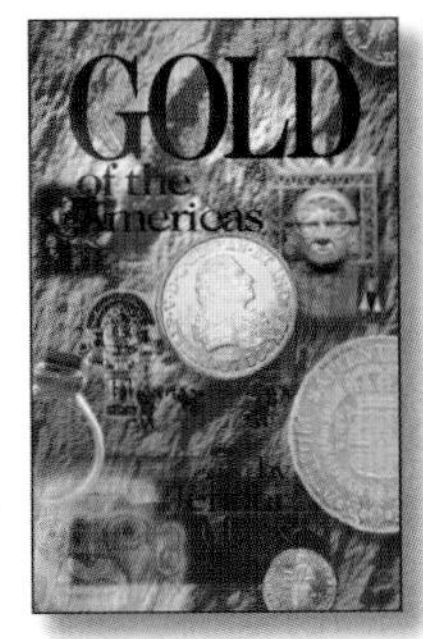

PN 1500300

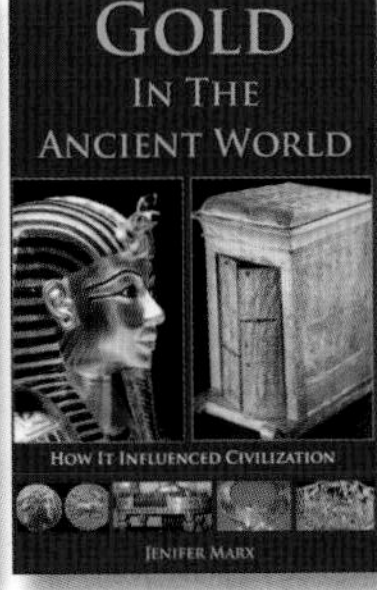

PN 1546100

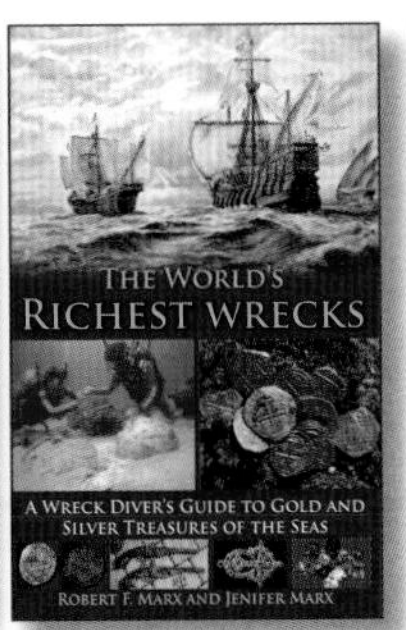

PN 1509810 hard cover
PN 1509800 soft cover

PN 1509900

TO ORDER RAM BOOKS

Please note that RAM Books, the publishing division of Garrett Metal Detectors, continues to release new titles each year related to treasure hunting, gold prospecting, coin hunting and relic recovery.

To see a current list of titles available from RAM Books, please consult a Garrett Metal Detectors hobby catalog or visit:

www.garrett.com

After reaching Garrett's website, visit the Hobby Divison section and select "RAM Books" to see all of our current titles. An order form is available on our site which can printed and mailed with your requested titles and payment.